Biocosmism

CRITICAL MEXICAN STUDIES

CRITICAL MEXICAN STUDIES
Series editor: Ignacio M. Sánchez Prado

Critical Mexican Studies is the first English-language, humanities-based, theoretically focused academic series devoted to the study of Mexico. The series is a space for innovative works in the humanities that focus on theoretical analysis, transdisciplinary interventions, and original conceptual framing.

Titles in the series:
The Restless Dead: Necrowriting and Disappropriation by Cristina Rivera Garza
History and Modern Media: A Personal Journey by John Mraz
Toxic Loves, Impossible Futures: Feminist Living as Resistance by Irmgard Emmelhainz
Drug Cartels Do Not Exist: Narcotrafficking in US and Mexican Culture by Oswaldo Zavala
Unlawful Violence: Mexican Law and Cultural Production by Rebecca Janzen
The Mexican Transpacific: Nikkei Writing, Visual Arts, and Performance by Ignacio López-Calvo
Monstrous Politics: Geography, Rights, and the Urban Revolution in Mexico City by Ben Gerlofs
Robo Sacer: Necroliberalism and Cyborg Resistance in Mexican and Chicanx Dystopias by David Dalton
Mexico, Interrupted: Labor, Idleness, and the Economic Imaginary of Independence by Sergio Gutiérrez Negrón
Serial Mexico: Storytelling across Media, from Nationhood to Now by Amy E. Wright
Sonic Strategies for Performing Modern Mexico by Christina Baker
Subjunctive Aesthetics: Mexican Cultural Production in the Era of Climate Change by Carolyn Fornoff
Fatefully, Faithfully Feminist: A Critical History of Women, Patriarchy and Mexican National Discourse by Carlos Monsiváis, translated and edited by Norma Klahn and Ilana Luna

Biocosmism
Vitality and the Utopian Imagination
in Postrevolutionary Mexico

Jorge Quintana Navarrete

VANDERBILT UNIVERSITY PRESS
Nashville, Tennessee

Copyright 2024 Vanderbilt University Press
All rights reserved
First printing 2024
Library of Congress Cataloging-in-Publication Data on file

ISBNs:
978-0-8265-0651-1 (Paperback)
978-0-8265-0652-8 (Hardcover)
978-0-8265-0653-5 (EPUB)
978-0-8265-0654-2 (Web PDF)

To María Elena González Borgaro
and Dylan Quintana González

Contents

Acknowledgments ix

INTRODUCTION 1

1 Alfonso L. Herrera's Plasmogeny:
The Inorganic Life of the Cosmos 21

2 Resurrecting the Past: Animality and
Chemical Ethics in Alfonso L. Herrera 60

3 José Vasconcelos: Botanical Ethics and
the Cosmic Race 97

4 Dr. Atl and Nahui Olin: Volcanism, Cosmological
Forces, and Space Exploration 136

EPILOGUE 179

Notes 185
References 211
Index 227

Acknowledgments

This book is the result of a long, often times unpredictable journey. When one finds themselves at the end of the process, it seems easy to ascribe a sort of necessary order or underlying logic to the whole journey. But the truth is that doing research and writing—as any other kind of vital activity—hinges upon a series of contingent encounters and small decisions, which end up having crucial, unintended consequences.

In retrospect, it is clear that the initial seeds of this project were planted during my time as a master's student at Universidad de Sonora, when I first became fascinated by postrevolutionary Mexican culture and started to explore some of its less-known facets. I'm forever indebted to all the faculty at Universidad de Sonora, especially to Fortino Corral, Rita Plancarte, and Gabriel Osuna, who taught me how to carefully read texts and supported me with overwhelming generosity. I was privileged to continue my graduate studies in the Department of Spanish and Portuguese at Princeton University, where I found a vibrant community that allowed me to deepen and broaden my intellectual interests. A special thank you to Rubén Gallo, Gabriela Nouzeilles, Rachel Price, Susana Draper and Pedro Meira Monteiro for encouraging me to see Latin American culture with fresh eyes and helping me in so many ways to develop as a young scholar. This book would not exist without their professional and intellectual mentorship and guidance. My hope is that this book showcases what I consider the main skills I have learned throughout my binational education: a rigorous, meticulous approach to analyzing texts (Mexican academia), and the intention of developing theoretically innovative readings of cultural production (US universities).

I was extremely fortunate to start my academic career as a faculty member in the Department of Spanish and Portuguese at Dartmouth College, where this book project started to take its current form. Since the moment I arrived at Hanover, I felt welcomed into the community and received the unwavering support of many colleagues: Silvia Spitta, Isabel Lozano, Israel Reyes, José del Pino, Rebecca Biron, Analola Santana, Julio Ariza, Natalia Monetti, Irasema Saucedo, Martina Broner, Samuel Carter, Beatriz Pastor, Annabel Martín, Mauricio Acuña, Ingrid Brioso Rieumont, Jorell Meléndez Badillo, Charlotte Bacon. I'm specially indebted to a group of colleagues who have become my dearest friends and mentors: Noelia Cirnigliaro, Sebastián Díaz Duhalde, Antonio Gómez López-Quiñones, and Carlos Cortez Minchillo. It is hard to overstate how much they have contributed to my personal and professional development since I got to Dartmouth. I would also like to extend my gratitude to the Leslie Center for the Humanities at Dartmouth, which hosted a seminar to discuss the first draft of my book. During this seminar, Ignacio Sánchez Prado, Rebecca Biron, Sebastián Díaz Duhalde, Mary Coffey, Adela Pineda Franco, and Antonio Gómez López-Quiñones provided invaluable insights and suggestions to improve my manuscript. Without their guidance I would not have been able to complete the book you are holding.

My research is also the indirect product of conversations and collaborations I've had with many friends and colleagues across the fields of Mexican and Latin American Studies. I am especially grateful to Jennifer Rodríguez, Pablo Domínguez Galbraith, Gerardo Muñoz, Susan Antebi, Horacio Legrás, Sergio Villalobos Ruminott, Jens Andermann, Gabriel Giorgi, Victoria Saramago, Ximena Briceño, Jennifer French, Samuel Steinberg, Carolyn Fornoff, Viviane Mahieux, Yanna Hadatty Mora, Elissa Rashkin, Rafael Mondragón, Amy E. Wright, Andrew Reynolds, Emily Hind, Pavel Andrade, Regina Pieck, Sergio Gutiérrez Negrón, Ana Sabau, Derek Beaudry, María Pape, Sebastián Figueroa, Cristóbal Jácome Moreno, Jesse Cohn, and Tania Aedo.

I extend my gratitude to Vanderbilt University Press, especially to series editor Ignacio Sánchez Prado for believing in this project and to editors Gianna Mosser and Zach Gresham for guiding me through the manuscript submission and publication process. My sincere thanks as well to Isis Sadek and Margaret Schroeder Urrutia for editing the manuscript at different stages and helping me improve my writing.

I would like to thank my family for their unconditional love and support: my parents, Miguel Quintana Tinoco and Elsa Navarrete Hinojosa; my brother, Miguel, his wife, Romina Molina, and their two children; my sister, Elsa, her husband, Javier Valenzuela, and their two children; my uncles and aunts, especially María de los Ángeles Navarrete and Sonia Quintana. It would be impossible to put into words how having a supportive family has shaped my personal and professional life in so many positive ways. A special thank you also to my dear friends from Sonora, Princeton, and the Upper Valley: Herlinda Jocobi, Alfonso Sánchez, Maribel Maldonado, Carlos Rascón, Luz Elena Meza, Adriana Velderrain, Samuel Paz, José Camacho, Miguel Carballo, Adrián Solórzano, Jorge Mondragón, Alfonso Ramos, Ulises Saldaña, Marco Gutiérrez, Leonardo Sánchez, Jonathan Aguirre, Christina Gonzalez-Aguirre, Brenda Bengoa, Carlos Castillejos, Alberto Rodríguez, Diana Marcela Rojas, Daniela Agusti, Felipe Severino, Karol Gomes, Luis Aníbal Gomes, and Francisco Moysés.

This book is dedicated to my wife, María Elena González Borgaro, and my son, Dylan. Their endless love and affection have filled my life with purpose and joy, carrying me through the challenging times.

Chapter 1 is a substantially expanded version of my article "Biopolítica y vida inorgánica: la plasmogenia de Alfonso Herrera," *Revista Hispánica Moderna* 72, no. 1 (2019): 79–95. A few paragraphs of the aforementioned article were also included in Chapter 2. Lastly, Chapter 3 is a revised and expanded version of my article "José Vasconcelos's Plant Theory: The Life of Plants, Botanical Ethics, and the Cosmic Race," *Hispanic Review* 89, no. 1 (2021): 69–92.

Introduction

In 1921, Diego Rivera was commissioned by Secretary of Public Education José Vasconcelos to paint a mural on the walls of the National Preparatory School auditorium. *La creación* became Rivera's first mural and one of the inaugural works of the renowned muralist movement in postrevolutionary Mexico. Inspired by Vasconcelos's philosophical ideas, Rivera decided to paint an allegorical account of the creation of the universe and its diverse manifestations, which included little or no reference to the Mexican Revolution and other nationalist motives. At the top of the mural, a blue semicircle filled with constellations represents what Rivera calls the "primary energy," a unique cosmic energy that has created and animated everything that exists.[1] From this semicircle the cosmic energy radiates in three main directions—indicated by three hands with pointing fingers—infusing with life orders of existence of increasing complexity and transcendence. In the mural's center, the cosmic vitality created what Rivera terms "the original Cell" surrounded by plant and animal beings in a natural environment that culminates in a nude human figure with open arms.[2] Humanity is thus introduced as part of a biological continuum starting with the first cell, which in turn stemmed from a cosmic energy that filled the universe. The open arms of the central human figure gesture toward the two sides of the mural, representing the masculine and feminine components of humanity. At the bottom, one can see a nude figure on each side—a woman on the left, a man on the right—embodying the corporeal and purely biological substrate of humanity, which acquires progressively higher faculties represented by allegorical figures ascending toward the primary energy. Ultimately, *La creación* depicts an image of the material creation of the universe

FIGURE I.1. Diego Rivera's *La creación*. Courtesy of Schalkwijk / Art Resource, NY, © 2024 Banco de México Diego Rivera Frida Kahlo Museums Trust, Mexico, D.F. / Artists Rights Society (ARS), New York

and the constitutive interrelation of everything that makes up the cosmos, from constellations to cells, plants, animals, and humanity.

As critics have emphasized, *La creación* does not feature the sociopolitical themes and revolutionary ideals that would become distinctive traits of Rivera's murals and of Mexican muralism in general during the next decades.[3] Thus, coinciding with Rivera's own assessment of this work, most critics have considered *La creación* as a preparatory, underachieving mural painting, a sort of initial attempt in which Rivera—and even Mexican muralism as a whole—was still developing the new pictorial and ideological approaches that would characterize his mature work.[4] Leonard Folgarait, for instance, deems this mural a "meeting of old and new" both in style and subject.[5] This view takes Rivera's subsequent murals—distinguished by a comprehensive depiction of Mexican society and history—as the teleological point of arrival of his mural painting, as if the characteristics of his most celebrated murals were an expression of the *true* and *necessary* character of his work and of the Mexican muralist movement. Conversely, I suggest thinking of *La creación* as an illustration of a set of ideas regarding the cosmos that coexisted—even if in a subordinated position—with the sociopolitical reflection and nation-building efforts undertaken by most murals and cultural artifacts of the time. In other words, Rivera's first mural is not a tentative, slightly misguided attempt at shaping postrevolutionary society and culture, but a full expression of certain biocosmic ideas overshadowed by more prominent cultural trends and themes. As I will show in this book, biocosmism formed a steady undercurrent that underlay philosophical, artistic, and scientific production in postrevolutionary Mexico.

Biocosmism was a term employed in intellectual circles around the world during the first half of the twentieth century. The International Biocosmic Association (Association Internationale Biocosmique) was founded in France during the 1920s by a handful of intellectuals, including Frenchmen Félix Monier and Albert Mary, Dominican Luis Arístides Fiallo Cabral, and Mexican Alfonso L. Herrera, one of the central figures of this book. But this association and its members were only one of many collectives and non-affiliated thinkers around the world who espoused a loose set of speculations concerning the cosmos, space exploration, and life extension.[6] While Herrera's connection to biocosmism is evident, other Mexican intellectuals studied in this book probably would not have considered themselves "biocosmists," and they had no known affiliations to any biocosmic associations. However, as Rivera's first mural shows, biocosmic themes appeared in a wide variety of cultural interventions in postrevolutionary Mexico—including in Rivera's later work, as I will show in Chapter 2, and in other artists predominantly associated with nationalist stances. Particularly, throughout the book I will focus on the thinking of Alfonso L. Herrera, José Vasconcelos, Dr. Atl, and Nahui Olin, while also showing how works by Alfonso Reyes, Antonio Caso, Frida Kahlo, and David Alfaro Siqueiros relate to the biocosmic framework.

Rivera's *La creación* illustrates vividly the general outlook shared by these intellectuals: they conceived of themselves not only as subjects of a national community, but more importantly as members of a biological species, as residents of the Earth and beyond. Their diversified set of activities at the intersection of art, science, and thought explored in various ways what they termed "universal life" or "cosmic vitality:" the idea that life animates and unites everything from cosmic space to inorganic matter to biological organisms. Vitality, according to these biocosmist intellectuals, is not a substance or faculty attached to a single kind of being: it is by definition a *cosmic* energy or process marked by continuity and interpenetration between the inorganic and the organic, human beings and nonhuman beings, the Earth and cosmic space. The notion that the cosmos is traversed by vital flows and displays a lively behavior had crucial implications for the ways in which biocosmists envisioned the current and, crucially, future relationships between the human species, nonhuman beings, and the universe as a whole. In this vein, embracing universal life opened up the space for imagining utopian projects such as the creation of synthetic life in the laboratory, the resurrection of extinct species, the conceptualization of botanical ethics, and the task of space colonization, to name just a few.

By tracing for the first time the intellectual constellation of biocosmism as it pertains to Mexico, this book explores a wide assortment of cultural interventions (philosophical, scientific, and literary texts, as well as visual artwork) touching on the vitality of the universe and its utopian potential, particularly during the postrevolutionary period in Mexico (1920–1940). This book delves into how a handful of Mexican intellectuals conceptualized the hitherto unexamined vitality of biological beings (cells, plants, animals), geological processes (inorganic matter, chemical reactions, volcanism), and cosmological entities (planets, outer space, galaxies), and whether this conceptualization reinforced or challenged the prevailing understanding of "the human" as the exceptional form of life. In other words, this book is about how these theorizations on nonhuman beings and processes related in conflictive ways with the anthropocentric and biopolitical frameworks of (human) community that underpin Western societies. One of the main arguments of this study is that biocosmic thought displays a constitutive, unresolvable tension or contradiction within itself: while it strives to produce a posthuman theory that embraces the agency of the whole universe, biocosmism often remains entrenched in the same anthropocentric, even eugenic assumptions that pervade the Western intellectual tradition. Rather than deeming this contradiction a shortcoming of biocosmist intellectuals, I maintain that it is a constitutive feature of the utopian task of imagining radical new ways of living and thinking: in order to open the door for radical novelty, human imagination is forced to take as a starting point the established ontological and epistemological notions with the ultimate hope of destabilizing them.

In addition to this conceptual or theoretical argument that runs through the book, *Biocosmism* also posits an interconnected, historical argument that relates to postrevolutionary Mexico. As I have already suggested, biocosmism is an intellectual trend that has been largely disregarded by cultural scholars who predominantly emphasize the importance of national identity and other sociopolitical themes during this historical period. In recent decades, there has been ample scholarly discussion of postrevolutionary theories of *mestizaje* and the biological traits of races, yet—as I will show later in this Introduction—these scholarly studies have largely overlooked concomitant theorizations of nonhuman life and processes. Throughout the book I argue that biocosmic thought simultaneously stemmed from and diverged from the well-known biopolitical reflection and praxis sustained in the wake of the Mexican Revolution. Biocosmism was embedded in the postrevolutionary emphasis on the biological aspects of human life, yet at the same time it strove to contest the anthropocentric premises of this general focus on life.

BIOCOSMISM IN MEXICO

In Mexico, biocosmic ideas thrived in the wake of the Revolution (1910–1917), which triggered a process of renovation that decisively transformed the face of the country. As Horacio Legrás puts it, "[h]istorians and cultural analysts agree that the revolution created modern Mexico and that very little in the way of a nation existed before Francisco Madero's call to arms."[7] Over the past decades, the humanistic and social disciplines have thoroughly studied the role played by thought and culture in the process of building modern Mexico during the postrevolutionary decades. In general terms, scholars have focused their attention on exploring issues such as modernity, political violence, gender, and race in relation to national identity and transnational circuits.[8] While these scholarly studies have greatly advanced our understanding of postrevolutionary culture and thought, the predominant critical emphasis on these identitarian issues does not exhaust the extraordinary complexity and richness of this cultural period. In fact, I argue that these primary areas of focus have precluded the examination of other areas of postrevolutionary culture that do not fit within this general scope, in the same way that the customary reading of Rivera's *La creación* has impeded a careful analysis of its biocosmic overtones. This is also the case for canonical works such as Vasconcelos's *La raza cósmica*, which has mostly been studied with an emphasis on its *racial* and not its *cosmic* aspects. In some other cases, the prevailing emphasis on national identity has overlooked ideas related to the natural world and biological life such as Herrera's, which have received scant scholarly attention within Mexican cultural studies. Even when scholars have delved into biocosmic texts of the postrevolutionary period, they have usually deemed them singular instances that touch on peripheral, unusual concerns. This assessment applies to the well-known cosmic ideas of Dr. Atl and Nahui Olin, which are often considered somewhat of an oddity to be studied as a separate case.

More than just a descriptive term, then, *biocosmism* is a heuristic tool that allows us to see postrevolutionary culture in a different light. As Carolyn Fornoff has recently argued for the case of Mexican contemporary poetry, a new field of enquiry opens up when the focus on the analysis of culture in relation to the national community is shifted to examining its relationship with the planet and the cosmos.⁹ This shift, as Fornoff puts is, "asks what happens when we stop equating Mexican literature (and culture) with the delimited confines of the nation-state" and ultimately allows us to explore "alternative geographies or timescales that are smaller, larger, or more scattered that the spatial-temporal bounds of the nation."¹⁰ In a similar vein, Cristina Rivera Garza has urged us to consider the geological or planetary dimension of literary works by theorizing the notion of *geological writings*: a poetics that *unearths*—in formal and/or thematic terms—the underlying strata composed of organic and inorganic beings and processes.¹¹ Rivera Garza's readings of the work of José Revueltas, Elena Garro, and other Mexican and Latin American writers has focused on bringing to the open their connection with a myriad of human and nonhuman agents overlooked by nationalist or humanist perspectives.¹² In particular, Rivera Garza's analysis of Revueltas's work, including the novel *El luto humano* (1943) and other texts, goes beyond the traditional political reading to explore issues closely related to biocosmic thought.¹³

In consonance with these and other planetary approaches to Mexican culture, one of the unique contributions of this book is the close examination of the multifarious work produced by Herrera under the name of plasmogeny, an intellectual project that encompassed a scientific theory of the origin of life on Earth and a theoretical reflection on life across the cosmos. Even if he is rarely included in studies of postrevolutionary thought and culture, Herrera does hold a prominent place in most histories of Mexican science, which establish him as one of the major figures in the professionalization of scientific research from the 1880s on.¹⁴ In particular, he played an important role during the fundamental transition from natural history to the modern paradigm of biological sciences and their institutionalization in Mexico.¹⁵ Herrera has also been recognized for introducing Darwinist ideas and perspectives into the country and, more specifically, for pioneering the global scientific research on the origin of life derived from a materialist, evolutionary framework.¹⁶ However, Herrera's philosophy of nature and his literary writings have hardly ever been analyzed before.¹⁷ By examining these largely unknown facets of his intellectual project, I hope to provide a more complex, multifaceted picture of one of the essential figures in the development of biological sciences in Mexico, while also enriching the cultural and philosophical archive of the postrevolutionary period.

Accordingly, this book devotes two chapters to exploring Herrera's broad range of scientific, artistic, and philosophical undertakings. Chapter 1 focuses on how he endeavored to prove his hypothesis on the origin of life by replicating its features in the laboratory. Understanding the material basis of life would, he thought, lead to a radical transformation of society, culminating in the artificial fabrication of a perfected humanity that would venture into space and colonize the universe. Herrera further developed a set of theoretical speculations that exceeded any biological definition of life by extending vitality to inorganic matter and the cosmos at large. Building on Alfonso Reyes's approach to plasmogeny in his essay "Amiba artificial" (1930), I interpret Herrera's experiments on the artificial creation of beings as an artform in itself. I argue that Herrera was interested in the artistic qualities of his experiments and the creative possibilities they opened up for perception and thought. Lastly, I analyze Herrera's poetry collection *Murmullos del universo*, published posthumously in 1982, as a poetic exploration of the notion of universal life.

Chapter 2 continues the examination of Herrera's work by focusing on his utopian projects for resurrecting bygone forms of life. First, I delve into his attempts to hybridize humans and apes in an attempt to artificially recreate the biological antecedent of humanity. In Herrera's mind, this experiment would prove the biological affinity between human beings and animals, underscoring the continuity of universal life and validating Darwinian, atheist theses. I critically examine how the project of hybridization was based on a reifying approach to animality and reinforced the racist assumption of the proximity between apes and non-Western races. The chapter subsequently focuses on Herrera's second plan for resuscitating forms of life: to identify and collect vestigial particles of past species with the aim of reconstructing their original form through technological means. I show how Herrera's mechanistic, teleological understanding of nature was staunchly criticized both by philosopher Antonio Caso and biologist Fernando Ocaranza in postrevolutionary Mexico. At the same time, the chapter demonstrates that the technological utopianism espoused by Herrera also appeared prominently in other biocosmic works such as Diego Rivera's *El hombre controlador del universo*. I argue that in the end, the mechanistic tendencies underpinning Herrera's plans for resurrecting lifeforms were paradoxically destabilized by his own notion of chemical ethics conceived of as the contingent, vital agency of inorganic matter across the cosmos.

Studying postrevolutionary culture in its relations with the cosmos—instead of the predominant focus on the role played by culture in defining national identity—not only opens up unexplored areas such as Herrera's work, it also provides the foundation for a rereading of canonical postrevolutionary texts such as José Vasconcelos's work. In contrast with Herrera, who was born more than ten years earlier than Vasconcelos and was intellectually educated in the scientific circles of late nineteenth century, Vasconcelos belonged to the generation of the Ateneo de la Juventud, a group of humanist intellectuals who dismantled the hegemony of positivist philosophy by introducing vitalist, spiritualist ideas drawn from the philosophies of Henri Bergson and Arthur Schopenhauer, among others.[18] Whereas Herrera arrived at biocosmic conceptions by radicalizing a materialist, scientific framework, Vasconcelos and his generational peers—including Caso, Reyes, and even Dr. Atl—rejected the idea that science should be deemed the only legitimate means of producing knowledge, and—without discarding altogether the data provided by the natural sciences—reappraised the transcendental importance of the study of literature, metaphysics, and the general realm of the human spirit. Even if he is widely considered a founding figure of twentieth-century Mexican philosophy, Vasconcelos's emphasis on spiritual realities has largely obscured the fact that his philosophical system builds on a full-fledged philosophy of nature and devotes several pages to speculating about nonhuman forms of life. In addition, his biocosmic speculations have also been disregarded by the critical emphasis on his reaffirmation of the *Mestizo* identity against hegemonic, Eurocentrist conceptions posed by Western thinkers. In this vein, historians of Mexican philosophy have mostly interpreted Vasconcelos and the Ateneo de la Juventud as philosophical predecessors of the "filosofía de lo mexicano," a line of thinking developed in the 1940s by the Grupo Hiperión—Emilio Uranga, Jorge Portilla, and Luis Villoro, among others—which set out to develop an original philosophy anchored on an analysis of the essential uniqueness of Mexican identity.[19]

In an endeavor to propose a new reading of Vasconcelos's philosophical works that destabilizes the prevalent interpretation, Chapter 3 explores his engagement with the notion of universal life and its implications for philosophical thought. Throughout the book I contend that the intellectual constellation of biocosmism was concerned with grasping and engaging in manifold ways with nonhuman and cosmological forces. Humans were thus conceived of as cosmic agents that penetrate and are penetrated by inorganic matter, other living beings, and the whole universe. Reading Vasconcelos as part of this intellectual constellation allows us to bring into consideration his speculations on animal and plant life, particularly his understudied theorization of botanical ethics. By drawing a comparison with Frida Kahlo's depiction of human–plant hybrids, I contend that both Kahlo and Vasconcelos reflected on the interpenetration and indistinction between plants, human beings, and the cosmos at large. In the last section, I examine how these philosophical ideas about plants provide the opportunity for an innovative reading of *La raza cósmica*. As I will show in this chapter, a majority of critical studies—mostly focused on the issue of race and the deconstruction of *mestizaje*—have disregarded the environmental implications and the strictly cosmic aspects of this utopia that I foreground in my analysis.

Chapter 4 concludes the study of biocosmism by centering on the cosmological works produced by Dr. Atl and Nahui Olin. These two figures are frequently studied together because of their intellectual and personal affinity, yet their biocosmic thinking has not hitherto been inscribed in a larger framework of related ideas. First, I examine Dr. Atl's understanding of volcanism as an expression of universal life through a reading of his literary work *Las sinfonías del Popocatépetl* (1921) and his scientific study *Cómo nace y crece un volcán: El Paricutín* (1943). I also take into consideration his landscape paintings, which used techniques—such as curvilinear perspective and the vibrancy of the "Atl colors"—that suggested a close engagement between the Earth and the cosmos. Then I delve into Nahui Olin's literary texts *Óptica cerebral* (1922) and *Energía cósmica* (1937) to expand on her notion of cosmic vitality as the harmonious movement of the totality. I particularly examine Nahui Olin's utopian desire to achieve a material transformation of the universe by exerting the power of the human brain and destroying the stability of the cosmic totality. I then focus on Dr. Atl's novel *Un hombre más allá del universo* (1935), which establishes a dialogue with controversial themes in the field of scientific cosmology. I analyze how this novel sets up a contradiction between the need to control and colonize the universe through space travel and the inexhaustible desire to discover the limits of the cosmos and human knowledge. Lastly, I examine the ways in which Dr. Atl's ideas culminated in the utopian project of Olinka, a city inhabited by an intellectual elite that would fulfil humanity's ideal of venturing into space.

Dr. Atl is a multifaceted figure who was simultaneously influenced by and contributed to the cultural and philosophical innovation introduced by the Ateneo de la Juventud.[20] He is predominantly known for his landscape paintings, which depict the natural environment of the Valley of Mexico and surrounding areas. Whereas these paintings are often analyzed by inscribing them in a nineteenth-century visual tradition of nationalist landscape art pursued by Eugenio Landesio and José María Velasco, among others, I propose in this chapter to frame them instead in connection with Dr. Atl's literary and scientific works on volcanism and cosmology, which, taken together, illustrate his biocosmic inclinations.[21] Some of Dr. Atl's speculations about the nature and behavior of the cosmos are radicalized by Nahui Olin's literary works, which—much like Dr. Atl's own novel—did not fit within the main trends of postrevolutionary literature represented by the so-called novel of the Revolution, the avant-garde poetry and narrative written by the *Contemporáneos* and *Estridentistas*, and other kinds of colonialist, proletarian, popular, or *cristero* narrative trends.[22] By emphasizing Nahui Olin's kinship with biocosmic ideas, I aim to show that her unique artistic participation in postrevolutionary culture can be productively examined in the context of shared notions concerning universal life.

Even though there was already a tradition of "cosmological discourse" in nineteenth-century Mexico, including Juan Nepomuceno Adorno's utopian thought, postrevolutionary biocosmic thought did not expressly mention it or take direct influence from it.[23] Instead, biocosmism drew on a wide variety of intellectual sources, from scientific research pertaining to biology and cosmology, to philosophical ideas related to materialism, vitalism, or esotericism, to artistic theories and trends. Mexican biocosmists were avid readers of innovative scientific literature and they themselves aspired to create new sciences such as Herrera's *plasmogenia* and Dr. Atl's *cerebrología*. They also thought that technological inventions could achieve unimaginable feats capable of transforming humanity's relationship with the cosmos. This "technological utopianism"—which was a broader feature of postrevolutionary culture, as Rubén Gallo has demonstrated—is illustrated by speculative mechanisms such as Herrera's microscope of darkness ("microscopio de las tinieblas") and Vasconcelos's electrochemical flows ("corrientes de electroquimia"), to name only two examples.[24] At the same time, biocosmists were drawn to exploring philosophical ideas that exceeded positivist paradigms: Bergson's vitalism played a crucial role in Vasconcelos's philosophy, the occultist ideas of Helena Blavatsky and astronomer Camille Flammarion were particularly important to Dr. Atl, and Herrera found the philosophical tradition of radical materialism significant. In addition to philosophy and science, experimental art constituted an important sphere of practice and reflection for biocosmists such as Dr. Atl and Nahui Olin. Art was construed as a suitable way to probe into the life of the cosmos and achieve a direct, expanded interaction with cosmological forces and inorganic matter.

Drawing on this wide array of sources, Mexican biocosmists engaged with science, philosophy, and art with the aim of imagining future arrangements of life. They were captivated by cutting-edge research and reflection conducted in each of those three areas, producing experimental forms of imagination that are not grounded solely in any one particular discipline. As a matter of fact, the process of professionalization and autonomization of science, philosophy, and art—to which biocosmists themselves contributed greatly—was still incipient during this time in Mexico. This situation originated a productive indistinction between disciplines that, far from being a deficiency or lack of professionalization, as some critics maintained during and after the postrevolutionary period, set the stage for the kind of experimental work and reflection conducted by biocosmists. In this manner, these intellectuals engaged with scientific, artistic, and philosophical notions in a way that established a creative tension between each of these fields as we conceive them today. For example, Herrera strove to showcase the aesthetically suggestive qualities of his laboratory experiments on the origin of life, a research topic deemed controversial and even illegitimate at the time, while also writing poetry full of scientific concepts and terminology that did not conform to contemporary literary trends. For their part, Dr. Atl and Nahui Olin relied on the most advanced research on cosmology—including contentious issues such as the shape, size, and configuration of the cosmos—to create their philosophically infused literary works, which also remained outside the mainstream of Mexican literature. Finally, steering away from Darwinist biology, Vasconcelos built on pioneering scientific research on plant life to develop his innovative philosophical theorization on ethics.

Radically interdisciplinary by nature, biocosmism occupies a mostly unacknowledged place in the historiographies of Mexican art, philosophy, and science, which tend to rely on a nationalist framework that focuses on how these fields have come to terms with national and foreign tendencies, thus contributing to the process of nation-building. By shedding new light on the intellectual constellation of biocosmism, I primarily aspire to make a contribution to the field of Latin American cultural studies with an emphasis on Mexico, the academic field in which I am formally trained and in which I regularly conduct research. Yet my hope is that scholars interested in the history of Latin American or Mexican science and philosophy might also find in this book new pathways for thinking through this historical period of renovation. My analysis will thus aim to bring to light how intensely scientific, philosophical, and artistic production intersected and fertilized each other during the postrevolutionary period.

BIOPOLITICS, LIFE, AND THE NONHUMAN

As Horacio Legrás has contended, the Mexican Revolution introduced a logic of social organization, which he calls *textuality*, that dismantled the hierarchical tendency of hegemonic powers by prescribing that the most contrasting social constituents—different gender identities, social classes, races, ethnicities, and so forth—deserved an equal footing in the national stage.[25] An elitist orientation that strived to imitate European civilization, oligarchical structures that favored an aged and mostly foreign economic elite, a social immobility that corralled citizens in the same geographical area and socioeconomic status in which they were born, the racial hierarchy that destined Indigenous peoples to the lower class—all these elements of Mexican society needed to be ushered out and reimagined. After years of armed struggle and the proclamation of a new Constitution in 1917, Mexico could not simply go back to its old ways; it had become something else, but it was difficult to say exactly what that was. For the next twenty years, the postrevolutionary regime endeavored to *produce* a revolutionary people, that is, to reinforce the values and attributes thought to be in accordance with the new times. In this process, as has been amply demonstrated by successive generations of scholars across the humanities and social sciences, arts and culture played a significant role by proposing a reinterpretation of national history from a revolutionary perspective and undertaking the reappraisal of popular and Indigenous traditions, among other projects.

In the last several decades, a great deal of scholarly attention has focused on how race—and related terms such as *mestizaje* or *indigenismo*—intersected with issues of national identity, modernity, and gender during the postrevolutionary period. Joshua Lund has argued that, beginning in the last two decades of the nineteenth century and especially after the Revolution, the Mexican state established itself as a "Mestizo state" that drew from the discourse of race to uphold its sovereignty and promote the penetration of capitalist accumulation, policing the distribution of material resources and political power along racial lines.[26] Similarly, Pedro Ángel Palou has delved into the postrevolutionary state's project to biopolitically control and optimize the racially diverse population by creating the hegemonic identity of the Mestizo.[27] David Dalton has contended that the postrevolutionary state fostered the use of various social and material technologies—modern machinery, industrial agriculture, eugenics, medicine, education, among others—to assimilate Indigenous bodies into a technologically advanced Mestizo identity, while Susan Antebi has explored how the state institutions strove to measure, document, and pathologize racialized, disabled bodies with the aim of shaping an eugenically "improved" individual and collective futures.[28] For their part, Mary Coffey and Adriana Zavala have described how the cultural project of *mestizaje* was a gendered undertaking that treated Indigenous and female bodies as naturally passive figures in clear contrast with the powerful depictions of masculine, Mestizo subjects.[29]

What these important contributions bring to light is the fact that postrevolutionary political culture fundamentally concerned itself with reckoning and tapping into the biological traits of (human) life. Mostly informed by biopolitical theory, these recent studies critically investigate what Foucault in his 1976 foundational take on biopolitics referred to as "the entry of phenomena peculiar to the life of the human species into the order of knowledge and power, into the sphere of political techniques."[30] As is well known, Foucault characterized biopolitics as a historically particular mode of exercising power that seeks to "*foster* life or *disallow* it to the point of death," in contrast with ancient sovereign power that strives to "*take* life or *let* live."[31] While traditional sovereignty fundamentally understood human beings as legal subjects who were exposed to being killed in compliance with the right of the sovereign, biopower assumes that human beings are living beings whose life must be administered and modified in conformity with desired norms. This "taking charge of life" by power implied that "the fact of living was no longer an inaccessible substrate that only emerged from time to time," a kind of stable phenomenon that power could simply take for granted as the inconsequential basis of the higher political realm.[32] Instead, life itself turns into a problematized, conflictive terrain that must be defined, studied, measured, and shaped accordingly. Rather than being a fixed essence, the life of individual and collective bodies is now conceived as a potentially unbounded, indefinite realm that needs to be contained and optimized by power-knowledge techniques such as statistics, eugenics, or public health.

As Foucault later explained that same year, one way in which biopower defined and framed the living characteristics of humans was precisely by "introducing a break into the domain of life that is under power's control: the break between what must live and what must die. The appearance within the biological continuum of the human race of races, the distinction among races, the hierarchy of races, the fact that certain races are described as good and that others, in contrast, are described as inferior: all this is a way of fragmenting the field of the biological that power controls."[33] In other words, the idea that the human species could be partitioned into different subspecies or races displaying specific biological traits provided the biopolitical mode of power with a foundation to decide which sections of the population should be valued, which ones were susceptible of improvement by adapting biologically, and which ones were destined to disappear or must be actively eliminated. In postrevolutionary Mexico, as in other Latin American countries, it has been shown that *mestizaje* was established as a eugenic mechanism for Indigenous people—deemed a lower form of life—to improve in biological and cultural terms, gradually approaching the highest type of human subspecies embodied by European White people.[34] This mechanism sought to create a whitened Mestizo population which would also be modern, literate, productive, healthy, and nationalist. Additionally, as I have argued elsewhere, *mestizaje* as a political technique also enabled implicitly the eugenic elimination of biological groups or races that were deemed too contrasting or deficient for them to adapt or "improve" biologically: this is the case of the Chinese and Chinese-Mexican population that suffered expulsion or death in postrevolutionary Mexico.[35]

Biocosmic thought examined in this book is embedded in the biopolitical horizon established firmly in the wake of the Revolution. These intellectuals' orientation toward the problematization of life informed their thinking and institutional endeavors in a profound way. At the outset, most of the biocosmist intellectuals studied here launched and/or supported the regime's biopolitical strategies to foster and optimize life in their capacity as state officials. For example, as Secretary of Public Education José Vasconcelos launched official campaigns for alphabetizing and assimilating Indigenous communities into capitalist society, while Director of the Directorate of Biological Studies Alfonso L. Herrera set out to survey and catalogue national biological resources with the purpose of promoting their exploitation. But more importantly, as I will show throughout the book, these biocosmic utopian undertakings relied on what Giorgio Agamben, who continued and transformed Foucault's approach to biopolitics, referred to as the biopolitical fracture between biological life (zoe) and politically valued life (bios).[36] These utopian projects operated by establishing a demarcation between "proper" and "improper" life, or between higher and lower forms of life within humanity and within the biological world at large. This becomes remarkably clear, as I will underscore in the following chapters, by examining the recurrent reliance on eugenic—or even Nazi in the case of Dr. Atl and Vasconcelos—ideologies as central components of biocosmic utopian imagination. Eugenic thinking—which was prevalent in Mexico and beyond during this time—informed the fashioning of these utopias of ultimate biological control that required the assumption of "humanity"—that is, the Occidentalized White or Mestizo human being—as the master and privileged resident of the universe.[37] Most of these projects posited an undisputed hierarchy of beings in which politically valued humans are granted the capacity to instrumentally manipulate and even recreate the human species and the biological world to "improve" them or adapt them to their needs.

At the same time, paradoxically, these utopian theorizations also contain ideas that call into question anthropocentric notions of human privilege and biopower. In this sense, while the recent contributions to the study of Mexican postrevolutionary culture have comprehensively examined the "biological-type caesura within a population," no critical attention has been given to how the shift toward the problematization of life simultaneously produced a timely reflection on nonhuman organisms and the characteristics of life removed from biopower's hold.[38] In other words, while recent studies have focused strictly on biopolitical divisions *within* humanity, this book endeavors to address this gap by examining philosophical, scientific, and artistic interventions regarding the unruly life of animals, plants, inorganic matter, cosmological forces. The overall problematization of what is life and of what it means to live thus informed the reconceptualization of other (nonhuman) forms of life, traditionally assigned to fixed, subordinate positions in the hierarchy of beings. If the biological life of humans was problematized as a contingent and indefinite realm, the vitality of nonhuman beings and processes was similarly reimagined to highlight their overlooked creativity and agentic capacities. In subscribing to the general logic of textuality (Legrás) predominant in postrevolutionary society, the central notion that the cosmos at large displays a vital behavior had the potential to disrupt the traditional hierarchies between "proper" and "improper" forms of life, between human beings and other biological organisms, between life and nonlife, and even between the Earth and other space bodies.

Thus, I contend that biocosmist thought simultaneously was embedded in and called into question the biopolitical framework espoused by the postrevolutionary regime and even by the biocosmists themselves. One could argue that the biopolitical problematization of life both *facilitated* and *restrained* the posthuman conceptualization of biological, geological, and cosmological processes. I will examine throughout the book how this fundamental tension or contradiction manifested itself in a variety of ways in biocosmic thought.[39] In some instances, biocosmists openly relied on dominant anthropocentric, biopolitical frameworks by suggesting the need to control and optimize nature and human beings alike, while in some other cases biocosmists themselves aspired to clear a pathway for an unconventional conceptualization of the agency of nonhuman agents and processes, which in turn required a renewed understanding of what makes humans *human*.

Even in these last instances, biocosmic thought often reverted back to anthropocentric premises, reproducing the same humanistic terminology—such as "consciousness" or "ethics"—that grounds human exceptionalism in the Western tradition. This seemingly inextricable link between anthropocentric and posthuman impulses shows how utopian imagination must necessarily grapple with current preconceptions and prevailing categories in its attempt to open a space for radical novelty.

In order to explore the biocosmic perspective on nonhuman beings and forces, this book establishes a productive dialogue with recent theoretical trends—new materialisms, posthumanism, as well as animal and plant studies—that go beyond biopolitical theory's focus on human beings.[40] These recent theoretical trends supplement biopolitical thought by stressing the fact that biopower's grasp on the living traits of human beings is founded on a previous assumption: what counts as humanity's life has been historically surmised by setting it against nonhuman living beings (animals, plants) and nonlife (inorganic matter).[41] In accordance with a long-standing intellectual tradition that stems from Aristotle's writings, biopower defines human life as rigorously opposed to the inertness of inorganic materiality, while also encompassing the vegetative and sensitive functions that pertain to plants and animals, respectively. At the same time, biopower suggests that the living traits of humanity can be controlled and optimized by means of the function of reasoning, which pertains exclusively to human beings and differentiates them from everything else. New materialisms, posthumanism, animal and plant studies explore critically what has been historically construed as the differences and similarities between human beings and other kinds of organic or inorganic entities, with the aim of reimagining the world ontologically. While biopolitical thought is concerned with mapping the effects of biopower's technologies and imagining ways of deactivating their hold, it does not endeavor to propose a new ontological understanding of plants, animals, and inorganic matter. In contrast, the aforementioned theoretical trends aspire to debunk the biological, political, and philosophical privileges ascribed to humanity in the Western intellectual tradition by reaffirming the overlooked agency and world-making capacities of materiality, animals, and plants.

These new theoretical trends thereby share two basic assumptions with biocosmism: on the one hand, the goal of challenging anthropocentrism or human exceptionalism, which frame human beings as *qualitatively* different from the rest of beings; and on the other hand, the view of nonhuman organic and inorganic beings not as passive objects of external forces that have a deterministic impact on them, but as inherently agentic or vital processes, engaging in deeply unpredictable and creative relationships with the world and between each other. These two theoretical tenets have received significant critiques—both from thinkers overtly against posthuman theories and generally attuned to them—that underscore what I have outlined as the lingering biopolitical, anthropocentric premises of biocosmism. For instance, it has been shown how vitalist new materialism—such as the one espoused in Jane Bennett's *Vibrant Matter*, for example—tends to *ontologize* the qualities historically assigned to "life," while also implicitly disallowing any activity related to "non-life" or passivity as valid ontological accounts of matter.[42] This ultimately leads to a so-called "flat ontology" based on the prescriptive claim that all human and nonhuman entities show the same potential agency and autonomy, which in turn might downplay the necessity of substantial political struggles against oppressive or exclusionary structures in modern biopolitical societies.[43] In sum, much like biocosmic thought, new materialisms tend to remain entrenched in hegemonic ontological and political premises in spite of their own effort to challenge those fundamental assumptions and imagine radically new ways of thinking.

Thus, establishing a dialogue between biocosmism and the recent theoretical trends allows me to achieve two interconnected goals. First, the emergence and consolidation of new materialisms, posthumanism, as well as animal and plant studies have granted the opportunity to look back and reappraise thinkers and trends such as biocosmism that remained neglected due to their marginal position. These recent theoretical trends provide a framework for making visible and studying collectively earlier ideas that were deemed eccentric or isolated. In this way, biocosmism can now be examined as a part of a long-lived tendency of thought that has continued until today and is currently thriving against the backdrop of the ecological crisis. Secondly, the study of biocosmic thought provides a magnifying glass for assessing both the deadlocks and potentialities of the current theorizations on nonhuman organic and inorganic beings and processes. As I argued earlier, both biocosmism and the most recent conceptualizations show a fundamental tension between their attempts to challenge human exceptionalism and the inadvertent reproduction of deep-seated categories underpinning the philosophical tradition of humanism and the consolidation of biopower. This fundamental tension is, in fact, the proof that the utopian imagination is at work, testing the absolute limits of hegemonic thinking with the hope of defamiliarizing it and rendering it less "natural" or "necessary." The utopian task of imagining a posthuman conceptualization and relationship with biological, geological, and cosmological processes captured the imagination of biocosmic thinkers at the beginning of the twentieth century and still continues to challenge us to find unpredictable ways of being and thinking.

UTOPIAN IMAGINATION

In his essay "El utopismo y la Revolución Mexicana" (Utopianism and the Mexican Revolution), historian Alan Knight concludes that "utopianism played a minor role in the Revolution" ("el utopismo jugó un papel menor en la Revolución") and that "in modern Mexico, as in England, the list of utopias is short, and the balance of utopianism, scarce" ("en el México moderno, como en Inglaterra, la lista de utopías es corta, y el saldo del utopismo, escaso").[44] Knight builds on a strict notion of utopianism that disregards utopian literature and focuses instead on "how utopian ideas and goals develop in historical processes" ("cómo las ideas y las metas utópicas se desarrollan en los procesos históricos") either in its "minimalist" (organization of independent and isolated communities) or "maximalist" versions (large projects and theories that seek to radically redesign society).[45] Regarding the latter version, Knight argues that the heterogeneous, contradictory ideology of the Revolution did not contribute to the formulation of coherent, all-embracing theories or projects; the closest case would be the anticlericalist Jacobinism of the 1920s that envisioned a secular reinvention of the Mexican people. Minimalist utopianism, although more akin to the contradictory nature of the revolutionary process, also failed to generate—according to Knight—a large number of experiences: the only indisputable case would be the colony of Santa María Auxiliadora, in Baja California, an intentional community founded in 1941 by the far-right movement Sinarquismo, which espoused a nationalist, messianic ideology in accordance with the social doctrine of the Catholic Church.

In contrast with Knight's perspective, I contend that mobilizing a broad theorization of utopianism as a "propensity" (Sargent), "impulse" (Jameson) or "desire" (Levitas) for future alternatives—which may vary significantly in terms of form, function, or content—provides a drastically different image of the postrevolutionary period.[46] At the outset, this theorization allows the possibility of studying a wide variety of utopian manifestations, including the so-called "three faces of utopianism"—utopian literature, intentional communities, and radical social theory—in postrevolutionary Mexico.[47] Projects of intentional communities, which involve a group of people who found an isolated community based on an alternative set of social guidelines, were far more common that Alan Knight concedes.[48] At the same time, utopian literature blossomed and produced an important corpus which for the most part has gained little scholarly attention, with the exception of Eduardo Urzaiz's well-known novel *Eugenia: Esbozo de costumbres futuras* (1919).[49] Social theory also took on a utopian tone during these times of unrest and constructive zeal.[50] In effect, one could argue that the set of uprisings and rebellions known as the Mexican Revolution brought about a destabilization of the well-established principles ordering and legitimating the social field in the Porfirio Díaz regime, which in turn triggered the desire and necessity of imagining new forms of existence to come. Once the hegemonic organization of discourses and bodies stopped making sense and crumbled, there was a sudden longing and practical need for envisioning unprecedented, radical ways of life.

The social theory of *mestizaje*, as developed during the postrevolutionary period, was perhaps the most preponderant form that utopian desire took during this period. By imagining the biological improvement of the population in the near future, intellectuals closely linked to political power such as Vasconcelos, Manuel Gamio, and Moisés Sáenz managed to "to turn [the state's] biopolitical control into a utopian enterprise" ("convertir su control biopolítico [del estado] en una empresa utópica").[51] As I have argued before, the intellectual constellation of biocosmism was fueled by the postrevolutionary turn to viewing life as a political issue, but it also exceeded this biopolitical framework in various ways. One could argue that the predominant trend of the postrevolutionary utopian desire was channeled through the theories of *mestizaje*, but there remained a surplus impulse irreducible to biopolitical, anthropocentric arrangements. This remainder opened up a utopian horizon not confined to proposing a biopolitical reconstruction of the nation-state that framed the territory's natural resources and the biological characteristics of the population as "national patrimony," but aspiring instead to transform the hegemonic relationships between human beings, nonhuman entities, and the cosmos at large. In the end, biocosmism tended to transgress the biopolitical matrix of the Mexican nation-state and sought to extend the utopian, anti-hierarchical impulse introduced by the revolutionary event to the whole universe.

Throughout the book I argue that biocosmism's utopian *impulse*—the inexhaustible drive to envision what is to come—often opposed and destabilize its own utopian *programs*, the specific set of principles guiding the projected alternative order.[52] While the utopian programs promulgated a specific "point of arrival" or desired outcome, the utopian impulse privileged the constant movement of the imagination guided by the criticism of the present and the hope for the future. Thus, the utopian impulse resonates with Miguel Abensour's notion of "the persistence of utopia," which "is not due so much to the repeated pursuit of a determinate content as to the ever-reborn movement toward something indeterminate."⁵³ In this way, insofar as the variety of programs espoused by biocosmic thinkers tended to reproduce biopolitical, anthropocentric assumptions in their images of the future, the underlying impulse displayed a propensity to criticize the established distributions of the world and open up the possibility of unimagined arrangements. In resisting solidification into an alternative program, the biocosmic utopian impulse introduced the permanent task of dismantling the biological, philosophical, and political hegemonic principles with the hope of clearing the way for future configurations of life.

CHAPTER 1
Alfonso L. Herrera's Plasmogeny

The Inorganic Life of the Cosmos

In his early work *Nociones de biología* (1904), Alfonso L. Herrera announced the creation of plasmogeny, a new science that, as its name suggests, centers on studying the genesis and constitution of protoplasm. Son of the renowned naturalist Alfonso Herrera Fernández, Herrera played a fundamental role in the institutionalization of biology in Mexico through his publications and his work in education and state-building.[1] *Nociones de biología*—the first book published in Mexico on what was still a young science at the time—showed that Herrera had read and assimilated the ideas of the founders of modern biology (Lamarck, Haeckel, Bernard, Darwin), while at the same time he had already started to develop his own theories concerning the origin of life. This book also showed that Herrera was willing to reach philosophical conclusions on topics such as the notion of life and the future of humanity. In addition to his publications, Herrera was in charge of presiding over the Dirección de Estudios Biológicos founded in 1915 by the Venustiano Carranza administration. During his term, which lasted until 1929 when this Dirección became part of the Universidad Nacional, Herrera promoted fundamental projects: he transformed the Museo Nacional de Historia Natural and founded the zoo, the botanical garden, and the aquarium, all located in the Chapultepec forest.

Throughout all these initiatives, Alfonso L. Herrera's vital and intellectual project was closely linked with plasmogeny as a discourse that aspired to have far-reaching implications in scientific, philosophical, and even artistic terms. Since 1910, he edited numerous publications on plasmogeny, including the pamphlet *Una ciencia nueva, la plasmogenia* (1911), *Biología y*

Plasmogenia (1924, an expanded edition of *Nociones de biología*), and his main scientific work *Una nueva ciencia: La plasmogenia* (1926, expanded edition of his 1911 pamphlet). Herrera also authored several philosophical writings like *Filosofía etérea* (1919) and one book of poems *Murmullos del universo*, which appeared posthumously in 1982. Herrera published profusely in scientific journals in Europe (France and Spain) and Latin America (Argentina and Cuba), reporting the results and implications of the experiments on plasmogeny that he conducted in the Dirección de Estudios Biológicos and, later, in a homemade laboratory. Herrera was also in charge of organizing congresses, scientific societies, and gazettes dedicated to plasmogeny, as well as an international network of scientists that included figures such as Jules Félix (Belgium), Israel Castellanos (Cuba), Víctor Delfino (Argentina), Albert Mary (France), among others. As the editors of an anthology and translation into English of some of Herrera's texts argue, "most of the foreign scientists with whom Herrera had strong academic ties were considered marginal in their respective countries" and "were in fact critics of the establishment with alternative ideas and proposals."[2] Plasmogeny was indeed a novel and innovative field at the time, so Herrera's experiments and theories on the origin of life caused admiration, disbelief, and even rejection. Today, despite the fact that his philosophical discourse and artistic production still remain unexplored, Herrera is recognized as a pioneering scientist in the investigation of the origin of life and in the area of synthetic biology.[3]

Plasmogeny was based on a theory developed in the 1860s by Thomas Graham and other scientists, who argued that protoplasm—the gelatinous substance found inside all cells—constituted the minimal substance that harbors life and, therefore, could be considered "the physical basis" of all existing forms of life.[4] Studying the genesis of protoplasm was, then, equivalent to offering a theory on the primordial origin of life on Earth. Darwin's *Origin of Species* (1859) had already suggested a preliminary hypothesis: that living beings could be thought of as the evolutionary result of an increasingly complex process of the organization of inert matter.[5] Derived from this general framework, Herrera's science was one of several theories on the origin of life that emerged at the time, including Alexander Oparin's theory, to which I will come back in Chapter 2. These scientific theories shared a rejection both of the ancient hypothesis of spontaneous generation (the idea of the sudden appearance of organisms lacking a process of gradual evolution) and the standard vitalist approach (which assumed the existence of a supernatural or divine force accounting for the vital functions of matter).

FIGURE 1.1. "Imitation of seeds, plants, sprouts, leaves. Metal salts in gelatin and potassium ferrocyanide," from Alfonso L. Herrera, *Una nueva ciencia: La plasmogenia*, 1926

Plasmogeny sought to verify its hypotheses by means of experimentation, that is, it sought to artificially imitate the protoplasm in the laboratory. Herrera and his colleagues carried out a large number of experiments that endeavored to reproduce the *form* of life, which according to their morphological viewpoint was the result of a close association between structure and function.[6] Thus, they strove to create not only the shape and structure, but also the vital functions of cells, tissues, organs, microorganisms, plants, and seeds. At the same time, plasmogenic experiments were conceived as heuristic tools that could bring about new hypotheses on the nature and origin of protoplasm and vitality at large.[7] The ultimate goal of plasmogeny was thus to achieve the synthesis of life in the laboratory. Starting in 1904 Herrera argued that "the imitations of the protoplasm are increasingly similar to the natural model, and those that are prepared with colloid silicates are almost equal to living matter, due to their structure and their power of absorption."[8] Later on, Herrera announced the creation of primitive organisms that he called "protobes," beings that displayed vital forms and functions and were thought to be the link between the mineral kingdom and the vegetable kingdom, between inorganic and organic matter.[9] In the so-called "protobial kingdom" one could find, for example, those pseudo-organisms baptized as "colpoids" and "sulphobes," which showed behaviors typical of life such as movement, irritability (changes experienced by external stimuli), nutrition (exchange of substances through osmosis), reproduction (duplication of an individual) and death (loss of impulses or movements after a certain period of time).

This chapter explores the political, philosophical, and artistic implications of plasmogeny. I will trace the fundamental contradiction at the heart of Herrera's intellectual project and of biocosmism at large. On the one hand, life was conceived as the object of biopolitical technologies that hierarchically divide desirable forms and residual forms. In this vein, encouraged by his success in artificially imitating vital features, Herrera imagined the utopian project of manufacturing a perfect race of "supermen" in the laboratory that would radically transform humanity's place in the cosmos. On the other hand, the discourse of plasmogeny paradoxically included a theoretical approach that tended to subvert such a eugenic project. I am referring to the notion of "universal life," according to which it is impossible to maintain qualitative differences between the animate and the inanimate or between the human being and other species, since all these categories turn out to be diverse manifestations of a single flow of inorganic matter that traverses the whole universe. I will also analyze how that same contradiction becomes apparent in the role played by Herrera in revolutionary governments, particularly in the conception of the Dirección de Estudios Biológicos and in the reformulation plan of the Museo Nacional de Historia Natural.

The second part of the chapter examines the artistic significance of Herrera's plasmogeny. The first section draws from the essay "Amiba artificial" (1930) by Alfonso Reyes, which is one of the most insightful explorations of what is at stake in Herrera's science. Reyes proposed that plasmogenic experiments could be considered a kind of artform in the sense that they aspire to *create* a new sensible reality. Exploring plasmogeny's relation to avant-garde poetry and contemporary bioart, I argue that Herrera maintained that art can be conceived of as a space in which life expresses itself: a type of inorganic life that is only distinguished by degree and not by substance from the life of organisms. In the second section, I argue that the poems compiled in *Murmullos del universo* stage a zone of undifferentiation between literary and scientific discourses that produces an innovative literary language. Paradoxically, this work thematically delves into the notion of universal life developed by Herrera, but at the same time it remains attached to a humanist conception of poetry and traditional formal strategies. This chapter not only highlights the unusual relevance of plasmogeny in the postrevolutionary context, but it also draws attention to the relationships of Herrera's thought with contemporary concerns in the fields of political, philosophical, and artistic reflection.

BIOPOLITICAL UTOPIA

In 1915, the engineer Pastor Rouaix—Secretary of Development, Colonization and Industry during the Carranza administration—commissioned Herrera to create a governmental dependency that "would give honor and prestige to our country in the scientific world, as well as internal profit and benefit with the perfect knowledge of its natural products and with the discovery of unknown latent riches" ("daría honra y prestigio a nuestra patria en el mundo científico, al mismo tiempo que provecho y beneficio interno con el conocimiento perfecto de sus productos naturales y con el descubrimiento de ignotas riquezas latentes").[10] It was a dependency fully inserted in the postrevolutionary project of inclusive modernization based both on the expansion of capitalism and the protection of the Indigenous masses through their political and racial incorporation.[11] The accumulation of capital was secured and expanded by the postrevolutionary state, which in turn based its legitimacy on the hegemonic construction of a national people.[12] In other words, the social process of capitalist accumulation required a biopolitical ordering of individual and collective bodies and of nature as resources awaiting to be exploited. In this context, one can easily understand the postrevolutionary drive to launch cultural and educational campaigns destined to "create" the revolutionary people: literate, hygienic, modern, and productive.[13]

This biopolitical project required an exhaustive knowledge of the territory, of the natural wealth and its relationship with the population that inhabited the territory. In his influential book *Forjando patria* (1916), Manuel Gamio highlighted the crucial political role that anthropology should play in this process, along with a variety of "complementary physiographic, biological, archaeological, historical and statistical-demographic studies" ("estudios fisiográficos, biológicos, arqueológicos, históricos y estadístico-demográficos complementarios") that is, a wide range of measurements, calculations and data that would enable governance.[14] According to Gamio, this scientific knowledge should inform the government's public policies, which would be aimed at promoting and optimizing the physical and moral health of the national "family," understood as a homogeneous group in political and racial terms. The revolutionary doctrine of *mestizaje*—in Gamio's work and more clearly in Vasconcelos's *La raza cósmica* (1925), analyzed in Chapter 3—was animated by the eugenic drive to form a biologically and culturally perfected people. Although the proposals of "mestizophilia" existed since the Porfiriato, the revolutionary regime adopted *mestizaje* as the basis for building a national people that would effectively integrate and neutralize the ethnic, socioeconomic, and political differences of the population.[15]

Herrera understood the relevance of the Directorate of Biological Studies in the context of the far-reaching biopolitical project of the revolutionary state. As announced in the opening speech of this agency, the Directorate would be organized into two interrelated entities: the National Museum of Natural History (the existing institution merged with the old Tacubaya Museum) and the Institute of General and Medical Biology (previously called National Medical Institute). The Institute would carry out the biological research that in turn would serve as the theoretical basis of both the applications (in medicine, agriculture, fishing, industry, etc.) and the efforts of exhibition and education of the masses (museums, talks, exhibitions, publications, etc.). In this way, the Directorate would fulfill three crucial functions: research, application and popularization/exhibition. Herrera was aware that, in the context of the transition between the Porfirian oligarchic regime and a new revolutionary social pact, the priority of the state could not be "purely theoretical studies" ("estudios puramente teóricos"), but rather it would be to find immediate public utility for the theoretical approaches of biology by exploiting the untapped natural resources of the country, "and later, with the product of those riches, founding establishments of theoretical interest" ("y más tarde, con el producto de esas riquezas, fundar los establecimientos de interés teórico").[16]

In response to the urgent needs of the country, the Directorate developed a series of fundamental initiatives. Firstly, the scientific exploration of certain regions (Oaxaca and Sonora, among others) was carried out with the aim of producing "biological charts" that recorded the existing natural resources and their possible uses in medicine, industry, and beyond. A study of infectious diseases caused by animals was also pursued: for example, a "Carta" dedicated to malaria wherein the mosquito that propagates this disease as well as the regions infested by this insect were biologically analyzed. The Directorate also contributed to manufacturing insecticides to combat pests such as the *gusano rosado*, which attacked cotton fields and reduced agricultural production. Furthermore, the Directorate provided technical knowledge to inform newly created regulations and laws that would ensure the conservation and efficient exploitation of forests and seas. In short, Herrera's work at the head of the Directorate was aimed at generating and applying scientific knowledge about the biological wealth of the nation and its use for the "improvement" of the population.

At the same time, Herrera considered that "plasmogeny represents the highest evolutionary degree of biology in our Homeland" ("la plasmogenia representa el grado más alto de la evolución de la biología en nuestra Patria") and, therefore, the Institute's research program dedicated the General Biology section to "the palpitating problems of morphogenesis, plasmogenesis and colloidal substances" ("los palpitantes problemas de la morfogenia, la plasmogenia y las sustancias coloidales").[17] The Bulletin of the Directorate for Biological Studies—eight issues of which appeared between October 1915 and March 1920—published regular articles on the most recent discoveries of plasmogeny. Encouraged by the utopian effervescence of the revolutionary decades, Herrera projected a hopeful future wherein the new science would be carried to its ultimate consequences. His optimism was based on the assumption that understanding the origin of the protoplasm would entail mastering the secret of life and its transformation to come:

> Medicine, once Plasmogeny has triumphed, will be able to cure or avoid all diseases, old age and death.
>
> Agriculture, livestock and, in general, animal husbandry, will be replaced by industrial food and materials, which will be artificial, just as is already the case with anilines, artificial silk, synthetic gasoline, rubber, etc. Human beings will devote their energy to intellectual work instead of plowing the fields and taming the foals or taking care of the chicken coops, becoming brute themselves in the process.
>
> Philosophy will extend life and its problems to the Universe and, perhaps, it and Science will be perfected by artificial men or brains.... Physical and moral pain will be extinguished forever, forever!
>
> *La Medicina, triunfando la Plasmogenia, podrá curar o evitar todas las enfermedades, la vejez y la muerte.*
>
> *La agricultura, ganadería y, en general, la zootecnia, serán sustituidas por los alimentos y materias primas de la industria, artificiales, como ya ocurre con las anilinas, la seda artificial, la gasolina sintética, el caucho, etc. Los hombres consagrarán su energía a labores intelectuales en vez de arar los campos y domar los potros o cuidar los gallineros embruteciéndose.*

> *La filosofía extenderá la vida y sus problemas al Universo y, tal vez, ella y la Ciencia serán perfeccionadas por hombres o cerebros artificiales. . . . Los dolores físicos y morales serán extinguidos para siempre, ¡para siempre!*[18]

Herrera and the plasmogenists hoped that the new discoveries would revolutionize all aspects of human society: nutrition, medicine, industry, education, and philosophy. Artificial foods, specially manufactured to satisfy the needs of the protoplasm, would protect the proper functioning of the base of life, "substituting one day not too far away the uncertainties of agriculture for plasmogenic organic chemistry" ("substituyendo un día no remoto la incierta agricultura por la química orgánica plasmogénica").[19] Understanding the physical and chemical forces that support the protoplasm would maintain health and even stop the aging process that leads to death. Herrera fantasized that plasmogeny would even make it possible to replace death "with a renewal similar to the phagocytic transformation of larvae, which change from caterpillars to butterflies by means of a profound renewal of their tissues" ("por una renovación parecida a la transformación fagocitaria de las larvas, que se cambian de orugas en mariposas por medio de una renovación profunda de sus tejidos").[20]

And not only physical phenomena would be dominated by plasmogeny, but also other aspects such as the psyche, learning, or thought. In an article published in the Bulletin of the Directorate, Herrera theorized "vibratory plasmogeny" as an ideal technique for imitating and perfecting mental faculties such as intelligence or memory. Building on the assumption that everything that exists is a product of the vibrations of the same basic matter, the Mexican biologist presumed that passions, feelings, and all mental phenomena should also be considered the product of these vibrations variously organized with particular "breadth, direction, number and period."[21] Herrera imagined that the invention of "vibratory mechanisms" composed of "sprockets, pulleys, motors, electric machines" would produce a large number and variety of vibrations, which could be optimized to achieve amazing results:

> Freeing the imagination completely free, just for a few moments, one can conceive of a thousand amazing wonders that future humanity will be able to achieve if it masters the problem of vibrations. . . . One can dream of vibrating devices that produce new vibrations in our nervous system accompanied by magical delights, and that by settling in the cortical structures of the brain, they would wonderfully increase the power of intelligence.

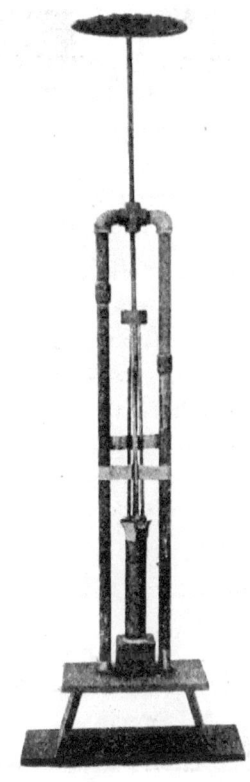

FIGURE 1.2. "Apparatus for injecting plasticine strings into human skulls," from Alfonso L. Herrera, *Una nueva ciencia: La plasmogenia*, 1926

Dejando libre a la imaginación completamente libre, sólo por algunos momentos, se conciben mil asombrosas maravillas que podrá realizar la humanidad futura si llega a dominar el problema de las vibraciones.... Puede soñarse en aparatos vibratorios que produzcan en nuestro sistema nervioso nuevas vibraciones acompañadas de mágicos deleites y que radicándose en las estructuras corticales del cerebro, aumenten de manera maravillosa el poder de la inteligencia.[22]

Herrera even undertook the task of artificially imitating the human brain, for which he built an enormous machine in the Directorate that injected a gelatinous substance into a human skull: "Since there are countless facts verifying the unity of the organs of our body, their common origin and the possibility of reproducing aspects, microscopic elements and even functions in the laboratory, we are right to generalize this possibility to all those organs, without excluding none of them, however complicated and surprising in their manifestations" ("Puesto que hay infinidad de hechos comprobando la unidad de los órganos de nuestro cuerpo, su común origen y la posibilidad de reproducir en el laboratorio aspecto, elementos microscópicos y aun funciones, tenemos razón al generalizar esta posibilidad a todos esos órganos, sin excluir a ninguno de ellos, por complicado que sea y sorprendente en sus manifestaciones").[23] Although the resulting "brain" was not (yet) capable of thinking, its striking morphological resemblance demonstrated that the characteristics of organic matter are not the product of an alleged life force. Herrera confessed his satisfaction at the scorn of his detractors: "This imitation of the organ of thought has awakened the fury of the orthodox people . . . and with that object in mind I undertook my work. It is beautiful to defy the storm. Experimental science sometimes allows itself these applications" ("Esta imitación del órgano del pensamiento ha despertado las furias de los ortodoxos . . . y con ese objeto emprendí mi trabajo. Es hermoso desafiar la tempestad. La ciencia experimental se permite algunas veces estas aplicaciones").[24] Once plasmogenic techniques were perfected, the human mind would reach unknown limits and artificial brains would be able to accumulate knowledge and think of the unimaginable.

THE HUMANITY TO COME

The maximum expression of the triumph of plasmogeny would unquestionably be the scientific creation of a new type of humanity. In 1911, the Mexican biologist had already maintained: "The theory of the Homunculus of Plato, Goethe and Hammerling, is no longer a utopia: before our eyes, very imperfect mineral figures of human embryos have been formed" ("La teoría del Homunculus de Platón, Goethe y Hammerling, no es ya una utopía: a nuestra vista se han formado uras minerales de embriones humanos, muy imperfectos").[25] Herrera tested various formulas to create these embryonic forms in his laboratory, and even took the time of subjecting the resulting embryos to an incubation process with the ultimate goal of developing human beings. In the future, the technique of synthetic embryos—eggs artificially fertilized by means of chemical substances, that is, without the participation of sperm—would be perfected to achieve the primordial desire of biologically improving the human species. Herrera's vision was radically optimistic: "In a more or less distant future, the eggs of new series of supermen will be prepared in laboratories, and the Earth will be populated with a new and perfect humanity, realizing the dream of Paradise, but of a Hellenic Paradise, where the supreme beauty of form, intelligence and virtue will reign" ("En un porvenir más o menos lejano se prepararán en los laboratorios los óvulos de nuevas series de los superhombres, y se llegará a poblar la Tierra con una humanidad nueva y perfecta, realizándose el ensueño del Paraíso, pero de un Paraíso helénico, donde reine la suprema belleza de la forma, de la inteligencia y de la virtud").[26]

In an article published in 1931, Herrera condensed his vision of the future humanity that would be produced by plasmogeny. After the confection of the simplest beings in the laboratory, plasmogeny would gradually increase its products' complexity until it would be able to synthetically create human beings:

> Later on, we will produce the metazoan and metaphites, the superior invertebrates, then the vertebrates and finally, a humanity, if one can call it like this, superior to the present one, without the disastrous atavisms of the beasts, morally and physically beautiful, Hellenic, of geniuses, artists, sages and philosophers, immortal entities, accumulating culture and gifts from centuries, migrating through space to other unknown worlds, freed from tyranny, pain, anguish, despair: with senses like the electric eye and the microphones, probing the eternal shadow from which everything comes, with the microscope of darkness, with the ether thermometer and barometer, luminous and perhaps intangible.
>
> *Más adelante se producirán los metazoarios y metafitos, los invertebrados superiores, luego los vertebrados y por fin, una humanidad, si así ha de llamarse, superior a la actual, sin los funestos atavismos de las fieras, moral y físicamente bella, helénica, de genios, artistas, sabios y filósofos, entes inmortales, acumulando la cultura y dones de los siglos, emigrando por el espacio a otros mundos desconocidos, libertados de la tiranía, el dolor, la angustia, la desesperanza: con sentidos como el ojo eléctrico y los micrófonos, auscultando la sombra eterna de donde viene todo, con el microscopio de las tinieblas, con el termómetro y el barómetro del éter, luminosos y tal vez intangibles.*[27]

This image of future humanity is largely predicated on a technological utopia. Future humanity will not only benefit from synthetic food, raw materials, and even artificial brains, it will also expand its faculties and potential through the invention of new machinic apparatuses. Just as in the case of the "vibratory mechanisms" enhancing human intelligence, Herrera envisioned that a novel set of machines—the "electric eye," the "microscope of darkness," the "ether barometer"—will enable surveying and controlling unexplored areas of nature: "probing the eternal shadow from which everything comes." Accordingly, future humanity will have the technological means necessary to move freely across the universe and even colonize the "unknown worlds" it might encounter. If restriction in space—living exclusively on Earth—will be overcome, the restriction in time—death—would also be remedied, since future human beings—these "immortal entities"—will achieve immortality by scientific means. Herrera's speculative technology would greatly amplify human perception and capacities, which in turn will secure humanity's mastery over the whole universe.

Plasmogeny's utopianism is based on framing nature as an external object that humanity can ultimately control and even rectify. In many ways, human beings *stand above* nature in plasmogeny's cosmovision. Similarly, Herrera's vision of future humanity implicitly puts forward a set of hierarchies and classifications within life. Herrera seeks to provide a better sense of the humanity he envisions by contrasting it with animals conceived as bare biological life: future human beings will live "without the disastrous atavisms of the beasts." If, according to Western thought, reasoning and culture set human beings apart from other animals, the future humanity envisioned by Herrera will thereby be composed of "geniuses, artists, sages and philosophers . . . accumulating culture and gifts from centuries." This future version of humanity will exclusively be devoted to intellectual endeavors. In this way, plasmogeny's utopian ideal also implicitly reinforces the Western hierarchy of mind over body: intellectual capabilities are privileged over bodily experiences, and intellectual work is certainly valued over manual labor in society.

Therefore, plasmogeny—or rather, a certain instrumentalization of it—reveals itself as a biopolitical technology that sought to organize the flow of life along divisions and hierarchies between desirable life and residual life, qualified life and bare life. The purpose of hierarchizing life is also clearly manifested in the diagram entitled "Evolution from ether to man."[28] It is a diagram that seeks to offer a comprehensive image of plasmogenic activity on planet Earth over time. At the lower end is the most basic matter (protyl or ether) that is organized above into elements and chemical compounds that in turn form the kingdom of protobes, from which two branches emerge: on the left side appear the "autotrophic bacteria" and, on the other side, the "pseudo-living proto-forms." This last branch undergoes a process of differentiation and complication that goes from the protozoans, passing through the moneras, the vertebrates and the fish until reaching the ancient anthropoids and the current human being, whose hierarchy rises from the "Blacks, Malays" to the "Mongols" and the "Yellows," culminating in the "Whites," which are explicitly presented as the superior form of civilization in the present. At the upper end of the diagram, above current

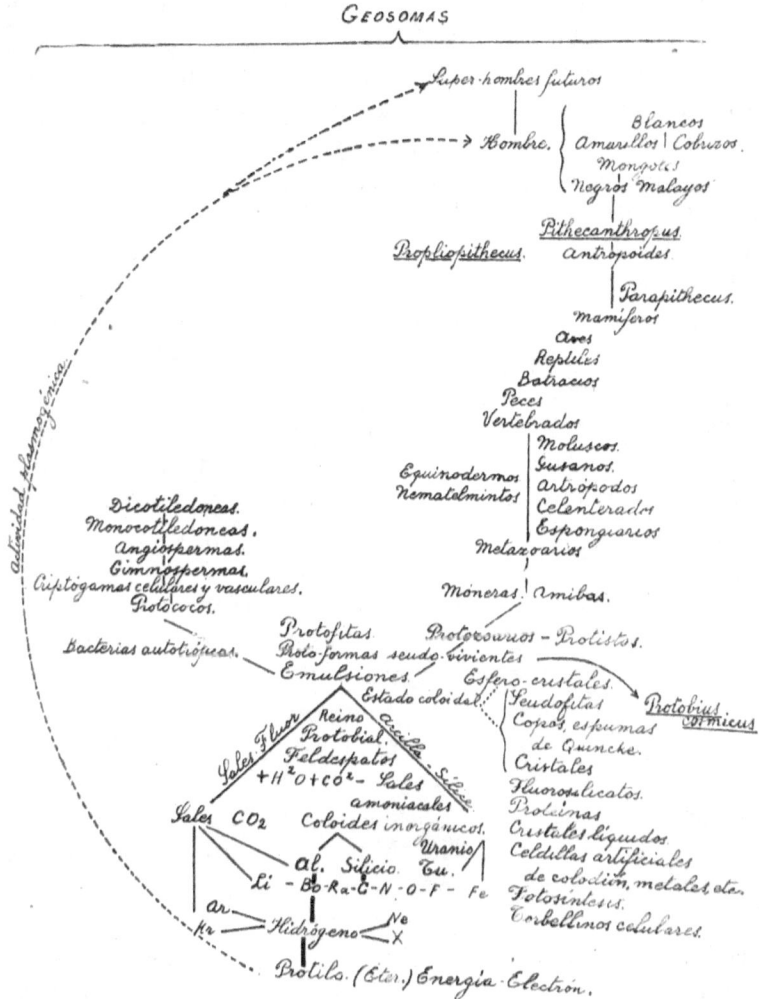

FIGURE 1.3. "Evolution from ether to men," from Alfonso L. Herrera, *Biología y plasmogenia*, 1924

humanity, we find the "future supermen" that would be created by plasmogenic methods in the not too distant future.

In this sense, it is revealing that, in offering the image of this future humanity, Herrera does not adhere to the postrevolutionary doctrine of eugenic *mestizaje*, but instead resorts to the qualifying term *Hellenic*, suggesting ancient Greece as the foundation of Western civilization: future humanity will thrive in "a Hellenic Paradise, where the supreme beauty of form, intelligence and virtue reigns."[29] Herrera involuntarily shows the "original nucleus" that biopolitical thought has also uncovered: the eugenic nucleus of sovereign power in the West from ancient Greece to the present day.[30] According to Agamben, the original structure that sustains power is based on the "inclusive exclusion" of *zoe*: the biological aspects of the human being must be excluded as unnecessary factors for the constitution of the political community, but through that same exclusion natural life ends up being "included" as the object of biopolitical mechanisms that try to manage and optimize it. This "original nucleus" will be reformulated over and over again by the state, the Church, and philosophy throughout Western history. This amounts to saying that politics, religion, and philosophy (metaphysics and even aesthetics), as they have been conceived by the mainstream of Western thought to this day, have functioned as what Sloterdijk calls "anthropotechnic," that is, technologies for the "production" of human beings through their deliberate breeding and domestication.[31]

During the first decades of the twentieth century, eugenics—the discipline founded by Francis Galton with the purpose of scientifically controlling human inheritance—became one of the preferred anthropotechnic methods for intellectual elites throughout the Western world. In postrevolutionary Mexico, there was no shortage of overtly eugenic initiatives, including Vasconcelos's proposal of an "eugenesia estética" in *La raza cósmica* and the founding of the Mexican Society of Eugenics for the Improvement of Race in 1931. In the Latin American context, some plasmogenists—such as Argentine Victor Delfino and Cuban Israel Castellanos—first introduced eugenic ideas and notions of "racial hygiene" in their respective countries. Castellanos makes explicit the connection between plasmogeny and eugenics: "artificial fertilizations, in a not too distant future, will be the substitutes for so-called eugenic marriages ... Herrera's science, as we see, will become an extremely effective auxiliary to Galton's science" ("las fecundaciones artificiales, en un porvenir no lejano, serán los sucedáneos de los llamados matrimonios eugénicos ... La ciencia de Herrera, según vemos, llegará a hacer una auxiliar eficacísima para la ciencia de Galton").[32] Plasmogeny and eugenics are presented, then, as two sister sciences that would guide humanity on its path of biological "improvement." Of course, both disciplines take for granted that Western cultural and physical standards define the esteemed ideal every individual and culture must aspire to.

Herrera's project of governing life in order to adjust it to eugenic standards is an undertaking that not only relates to the postrevolutionary government's attempt at physically and morally "redeeming" the Indigenous people, but also constitutes a program that is largely preserved in our time.

After protoplasm was replaced by the gene as the substance that purportedly harbors the nucleus of life, in recent decades there has been a proliferation of biotechnologies that focus on genetic material: from the possibility of altering genes in order to prevent diseases, to the manipulation of human DNA for the purpose of correcting genetic defects, to genetically modified food and DNA databases as biometric methods of identification and surveillance. All of these purposes—which echo the hopes harbored by plasmogenists in the early twentieth century—lead to the idea of manufacturing an anthropological device that could produce a higher standard of *the human*. Ultimately, the critical examination of plasmogeny shows that the foundational structure of Western thought operates on a eugenic rationality that presents itself as *natural* and universal, while being in fact an unnatural model imposed on life.[33]

However, as we will see next, the discourse of plasmogeny also hosts an impulse that, paradoxically, tends to call into question the eugenic rationality that anchors Western anthropological devices. Herrera's reflection shows that, although the constitutive plasticity of life provides the opportunity to control and codify it according to naturalized hierarchies, that same plasticity simultaneously tends to interrupt imposed, normative models. Biocosmic thought manifests itself, then, as an ambivalent terrain where discordant impulses enter into conflict: on the one hand, the purpose of directing life according to previously established, hierarchical criteria and, on the other hand, the constant endurance of the flow of life. This vital becoming—conceptualized as a universal, inorganic flow—guided both the philosophical discourse of plasmogeny and some of Herrera's institutional initiatives.

UNIVERSAL LIFE

In addition to endorsing the biopolitical project of the postrevolutionary state, the Directorate of Biological Studies had theoretical aspirations that, in Herrera's opinion, made it a leading research center in the world. The fundamental theoretical underpinnings were destroying the "idols of vitalism and animism" ("idolillos del vitalismo y animism"), proposing a holistic philosophy of nature, and showing that "everything is life where everything is movement, from the nebula to the infusory" ("todo es vida donde todo es movimiento, desde la nebulosa hasta el infusorio").[34] In this context, one of Herrera's most cherished institutional projects was the foundation of a Natural History Museum that would follow the modern notions of biology. In 1895, Herrera had published in French the article "Les musées de l'avenir," wherein he developed a theoretical proposal for the organization of new biological museums. While traditional natural history museums focused exclusively on classifying living things into categories such as genera, species, or classes, museums of the future would be organized based on a "philosophical classification" to reveal the inner workings of nature (principles such as the unity of nature, inheritance, variation, etc.). The museums of the future will not be "shops for things killed and eaten by worms" but "open books in which you can read the philosophy of nature."[35] In sum, instead of preserving the merely taxonomic zeal that seeks to immobilize and isolate life, the new museums would embrace the biological theories of Darwin, Haeckel and others, with the ultimate aim of offering a synthetic and dynamic understanding of nature and living beings.

Almost two decades after this article was originally published, Herrera had the opportunity to partially apply his ideas on the new biological museums at the National Museum of Natural History. This institution was organized into four sections: the first three (zoology, botany, geology/mineralogy) followed traditional taxonomic criteria, while the last section on biology (named "New Museum" by Herrera) adopted certain proposals on the museums of the future. This last section not only included "the objective demonstration of Darwin's theory of evolution (Cabinet No. 253)", but also incorporated a "showcase of Plasmogeny (No. 264), science constituted in Mexico" which exhibited the results of the artificial imitation of life (cells, brains, protobes, etc.) and an illustration of the plasmogenic theory on life's origin.[36] Altogether, in addition to exhibiting the "incalculable wealth of the Mexican Republic" and serving as a "center for popular education and recreation," this Museum showcased certain crucial theoretical notions, including the unity of nature, Darwin's theory of evolution, and the inorganic origin of life.

Thus, in addition to fulfilling a "patriotic and technical role" for public benefit, the Museum had theoretical objectives guided by Herrera's broader philosophy of biology. In particular, by demonstrating biological principles such as natural unity, evolution, and the inorganic origin of life, the last section of the Museum showed how the same basic matter constitutes all inorganic and organic bodies through dynamic processes of organization, variation, and transformation. The Museum thus carried out a double task of intellectual propagation: it sought to increase technical knowledge in order to overcome the general "apathy" toward the exploitation of the country's "incalculable wealth," but more importantly it also endeavored to teach the skills of "comparing, contemplating and inducing" with the aim of approaching nature with a holistic, undogmatic perspective.[37] In this sense, the conceptual design of the Museum was based on a theoretical reflection derived directly from countless experiments performed in the laboratory. According to Herrera, drawing from Haeckel's theories, plasmogenic experiments scientifically demonstrated that there is no essential difference between inorganic/inanimate and organic/animate matter: on the contrary, both categories of matter are distinguished by a mere difference in degree or combination, not by a qualitatively different substance or impulse.[38] In accordance with other biocosmist intellectuals studied in this book, he concluded thereby that the universe is made up of the same flow of matter that through a process of continuous transformation generates all the existing phenomena, from planets to simple organisms to human beings. Herrera summarized this notion in "the law of universal life":

> There is no living matter and no dead matter, because everything lives in the Universe; everything that exists is reduced to the mass or quantity of matter that a body contains, and its life, which is movement; the microcosm or small world, atoms, molecules, is an imitation of the macrocosm or great world and a chain of union and causality unites the nebula origin of the Sun and the planetary system, to the protoplasmic base of the organism, because life extends from the simpler being to the constellations of the Zodiac, and not enough distinction has been found between the living and the inert.

> *No hay una materia viva y una materia muerta, porque todo vive en el Universo; todo lo que existe se reduce a la masa o cantidad de materia que contiene un cuerpo, y su vida, que es el movimiento; el microcosmos o pequeño mundo, átomos, moléculas, es una imitación del macrocosmos o gran mundo y una cadena de unión y de causalidad une la nebulosa origen del sol y del sistema planetario, a la base protoplásmica del organismo, porque la vida se extiende del ser más sencillo a las constelaciones del Zodiaco, y no se ha encontrado una distinción suficiente entre lo vivo y lo inerte.*[39]

The notion of universal life deconstructs the traditional division between the living and the nonliving. The movement that animates everything that exists in the universe has various manifestations, among which is biological life defined as "the physical-chemical activity of the protoplasm."[40] But plasmogenic experiments show that it is a mistake to grant a special status to the life of biological forms, for ultimately they are nothing more than a specific kind of organization of universal matter. In fact, living organisms strictly depend on inorganic/inanimate matter to the extent that they could not exist without the universal circulation of energy and matter. As Herrera points out, "a *chain of union and causality* unites the nebula origin of the Sun and the planetary system, to the protoplasmic base of the organism."[41] There is a causal relation between our planetary system and living organisms because biological life is the result of a complicated process of evolution that started when the sun and all the planets were first created. The same chemical elements and physical forces that originated the solar system were the natural agents that eventually produced living organisms on Earth. And these lifeforms continue to depend on cosmic forces such as sunlight that, once is converted into chemical energy through photosynthesis, supplies most of the energy required to sustain life on Earth. In this sense, in addition to its cosmic origin, biological life is still closely interrelated with the inorganic/inanimate matter and energy that populate the universe.

The notion of universal life, then, calls into question the assumption—a founding premise of biology as a modern science—that living organisms are self-contained entities that possess a special function called "life." Rather, the idea that "everything lives in the Universe" exceeds any strictly biological definition of life and proposes, instead, that life is a cosmic process of the movement of matter. Herrera even goes beyond to suggest that not only biological life is simply a manifestation of universal forces, but also the universe itself is conceivably populated by living organisms. Herrera rejects what he calls the "biocentric error": supposing that biological life is attainable exclusively on a single planet in the whole universe. He does not believe that the Earth has a special nature that sets it apart from the rest of the universe. On the contrary, as he contends: "The existence of the same elements in the various celestial bodies, according to Lockyer, and the universality of the laws of nature, lead us to think that life exists everywhere" ("la existencia de los mismos elementos en los diversos cuerpos celestes, según Locyer, y la universalidad de las leyes de la naturaleza, nos inclinan a pensar que la vida existe en todas partes").[42] Herrera even proposes the "law of planetary lives," according to which biological life will be produced every time that a planet produces the necessary "physical and chemical conditions" to foster protoplasm.[43] In those planets, which might already exist in some part of the universe, living organisms would go through the same process of evolution that marked the history of life on Earth, eventually leading to the existence of human beings and maybe "other beings of very complex and even more perfect organization than that of humans" ("otros seres de organización muy complexa y aun más perfecta que la del hombre").[44] Viewed in this context, it makes sense that Herrera assumed that the future humanity produced by plasmogeny would have to engage in space exploration and colonization.

By understanding life as a process involving all inorganic matter in the cosmos, the notion of universal life tends to disrupt the hierarchical relationship of human beings toward other forms of life and the natural world. In this sense, the idea of universal life does not only imply the deconstruction of the traditional opposition between the living and the nonliving, but also involves relativizing the divisions and hierarchies that have organized biological life in Western societies since Classic antiquity. From the outset, Herrera disapproves of the abstract categories such as "species" that try to immobilize and standardize the singularity of all beings, because "in nature there are only individuals" ("en la naturaleza solo hay individuos"), that is, there are only particular beings in constant transmutation.[45] The Mexican biologist also argues that it is necessary to reject "that absurd theory of the impossible isolation of our species, which should only be related to divinity" ("esa teoría absurda del aislamiento imposible de nuestra especie, que sólo debía estar relacionada con la divinidad"): the assumption that human beings have an essential character that make them superior to other species and intrinsically related to God.[46] On the contrary, the conjecture that other kinds of humanities probably live in some part of the universe undermines any sense of fundamental uniqueness attributed to human beings. Furthermore, just as the other beings existing on Earth, humanity should be conceived of as a different modality of organization of the same cosmic matter that makes up everything that exists: "Our species and the universe breathe each other, like the animal and the plant, giving and taking oxygen and carbonic acid. The matter of our flesh comes from the planetary system and this from the universal system, which rotates on the cog of the Infinite" ("Nuestra especie y el universo se respiran mutuamente, como el animal y la planta, dando y tomando oxígeno y ácido carbónico. La materia de nuestra carne es del sistema planetario y éste del sistema universal, que gira sobre el piñón del Infinito").[47] This ultimately leaves no room for exalting the allegedly exceptional character of humanity.

INORGANIC LIFE

In a similar way, Herrera endeavors to relativize the traditional hierarchy of life by showing that the most rudimentary of organisms—an even inorganic matter itself—possess some of the high characteristics associated with humans, such as awareness and intelligence. For Herrera "awareness" is the name of the physical and chemical associations and dissociations that all kinds of matter constantly engage in: "all matter, animate or inanimate, alive or dead, reacting with the environment, is aware that it exists.

The monolith that rolls through the snowdrift, the acid that slumbers in the flask, the atom that vibrates in the inner night, are aware of themselves" ("toda la materia, animada o inanimada, viva o muerta, reaccionando con el medio, tiene la conciencia de que existe. El monolito que rueda por el ventisquero, el ácido que dormita en la redoma, el átomo que vibra en la noche interior, tienen conciencia de un 'yo'").[48] Defined as ubiquitous and incessant reactions, awareness is one of the essential traits of universal life and it tends to grant the same ontological dignity to all existing beings, including to those "colpoids" or rudimentary life forms fabricated in the laboratory. Herrera meticulously describes the tumultuous behavior of these curious beings, highlighting their intelligence and intentionality:

> From all points of the preparation globules sprout, which come from the segmentation of the drops, or amiboid figures of very different shape and size, which throw themselves at each other, as if possessed of an indescribable frenzy, and when touched by surfaces *exert a very active, mutual, violent suction.* . . . Something seems to awaken in them a desire or will, or rudimentary tact (tactism), a primitive intelligence, and again they throw one on the other and paralyze again. . . . On numerous occasions it is seen that the smallest, most agile, attack the big ones, collide against them, bounce like a ball, attack again, and come and go around a large Colpoid, as if recognizing it, looking for the most vulnerable, to fit their teeth, in the same way that a mosquito or flea runs on the hand looking for the weakest part to bite us, or the most tasty and rich in blood. This is when, above all, we realize that the Colpoid is not an inert globule simply dragged by osmotic currents or that it falls in one direction or another due to differences in level or accidental currents. No, obviously, he knows what he is doing, he is going after a determinate objective, he experiments, like the amoebas studied by Lillie, who follow a plan to escape from somewhere. . . . It is impossible to resist believing that *they have, therefore, a rudimentary consciousness or faculty to know.*

> *De todos los puntos de la preparación brotan glóbulos, que provienen de la segmentación de las gotas, o figuras amiboideas de muy distinta forma y tamaño, que se arrojan unas sobre otras, como poseídas de un frenesí indescriptible, y al tocarse por las superficies de contacto ejercen una succión activísima, mutua, violenta. . . . Algo parece despertar en ellos como un deseo o voluntad, o tacto rudimental (tactismo), una inteligencia primitiva, y nuevamente se arrojan uno sobre el otro y vuelven a paralizarse. . . . En numerosas ocasiones se ve que los pequeños, más ágiles, embisten a las grandes, chocan contra ellos, rebotan como pelota, vuelven a atacar, y van y vienen alrededor de un Colpoide grande, como*

reconociéndolo, buscando los puntos más vulnerables, para encajarles el diente, de la misma manera que un mosquito o pulga corre sobre la mano buscando la parte más débil para picarnos, o la más sabrosa y rica en sangre. Aquí es donde, sobre todo, nos damos cuenta de que el Colpoide no es un glóbulo inerte arrastrado simplemente por las corrientes osmóticas o que cae en un sentido u otro por diferencias de nivel o corrientes accidentales. No, evidentemente, sabe lo que hace, se dirige a un fin determinado, ensaya, como las amibas estudiadas por Lillie, que siguen un plan para salir de algún sitio. . . . Imposible es resistirse a creer que tienen, por tanto, una conciencia rudimentaria o facultad de conocer.[49]

Most scientists would argue that these elemental beings do not act purposely, and they simply behave following external conditions or a very basic instinct. But Herrera is prepared to assert that these colpoids already display signs of volition, intelligence, understanding, awareness, and consciousness. This is why, according to Herrera, plasmogeny "demonstrates that inferior beings reason, generalize, compare, invent, feel, love, hate and evolve intellectually as we do, with differences of degree but not of essential character" ("demuestra que los seres inferiores razonan, generalizan, comparan, inventan, sienten, aman, odian y evolucionan intelectualmente como nosotros, con diferencias de grado pero no esenciales").[50] Being constituted by the same cosmic matter and following the same natural laws, it is unreasonable to presume that the various kinds of beings are distinguished from each other by an *essential* difference. Colpoids' behavior proves that intelligence and reason are not a product of a uniquely human force, but a result of chemical reactions that can be found all across the universe. Of course, humanity's intelligence functions at a higher degree of discernment, but this fact should not be used to sustain the illusion of human exceptionality.

Humans, colpoids, and all other beings share a fundamental nature that is more significant than their differences. All beings are endowed with universal life and are "brothers in the great family of the Cosmos" ("hermanos en la gran familia del Cosmos").[51] This is what Herrera calls "universal fraternity" and was also theorized as "biocosmic solidarity" by members of the International Biocosmic Association.[52] This notion asserts that, rather than aspiring to stand above nature and control it, human beings must understand their existence as a minimum component of universal processes of movement, evolution, struggle, association, and so on. Embracing universal life "tends to give to man a new ideal, greater and more comforting than all of them: the blending of the individual in the whole, participating in the universal existence, within the unity of nature" ("tiende a dar al hombre un nuevo ideal: compenetración del individuo en el conjunto, participando de la universal existencia, dentro de la unidad de la naturaleza").[53] In this way, biocosmic solidarity calls into question Herrera's own assumption that humanity could harness natural laws and even substitute nature for an artificial "second nature" once plasmogeny were perfected.[54]

Deconstructing the privilege attributed to humans over all other kinds of beings and more generally to the living over the nonliving, Herrera's notion of universal life can be understood as an inorganic and anarchic life that disrupts traditional ontological categories and divisions in the Western world. Just like Herrera, Deleuze and Guattari developed their notion of "inorganic life" under the assumption that in nature "everything is alive":

> [it] liberates a power of life that human beings had rectified and organisms had confined, and which matter now expresses as the trait, flow or impulse traversing it. If everything is alive, it is not because everything is organic or organized, but, on the contrary, because the organism is a diversion of life. In short, the life in question is inorganic, germinal, and intensive, a powerful life without organs, a body that is all the more alive for having no organs, everything that passes *between* organisms.[55]

According to Deleuze and Guattari, inorganic life has the characteristic of constantly flowing and not allowing itself to ever be confined in an organized space. It is, therefore, a becoming without a subject (it is not the life *of* something or someone): an immanent movement lacking a final, teleological objective or horizon. Conceived in this way, life is "made up of virtualities": sometimes it is actualized but it is never fully exhausted, and it takes specific forms without limiting itself to one of them.[56] The stable hierarchies between the living and the inert, between the human and the nonhuman or between bare life and political life are called into question when taking into account this universal, inorganic flow that traverses all dichotomies and opens up a space for new configurations of life.

The notion of universal life is a philosophical wager that paradoxically deterritorializes the eugenic instrumentalization of plasmogeny. Herrera's speculations offer a possibility of deactivating the eugenic rationality of Western societies by proposing a definition of life as an inorganic and anarchic flow that destroys biopolitical norms and hierarchies. Life as "pure virtuality" is inexhaustible and ungovernable: it tended to subvert the biopolitical project of postrevolutionary governments, but more generally it also strives to destabilize the founding dichotomies of Western thought and politics. Herrera and plasmogeny thereby offer us a utopian formulation of the immanent movement of life and its constant becomings.

THE ARTISTIC IMAGINATION OF ALFONSO REYES

In 1930, Alfonso Reyes published "Amiba artificial" in the journal *Contemporáneos*. This short essay is one of the most significant meditations on the scientific, philosophical, and artistic implications of plasmogeny. Even though Reyes is a canonical writer whose work has been thoroughly studied, his writings related to science are mostly understudied. As an anthology on this topic shows, Reyes was well-versed in recent scientific theories and interested in exploring research fields as diverse as physics, biology, chemistry, among others.[57] Reyes's essays concerning science are usually speculative, ludic pieces that reflect on the philosophical and cultural implications of scientific thought. More than simply presenting scientific theories, these essays seek to explore how the elucubrations of science can spark unexpected reflections and imaginative associations within other human realms. The fact that "Amiba artificial" appeared in *Contemporáneos*—a journal led by the group of innovative poets named after it—already signals its general intent: bringing scientific thought to bear on art and cultural imagination at large. At the same time, Reyes draws on scientific theories to shed light on philosophical issues such as vitalism, which was crucial for his own intellectual project and for other members of the Ateneo de la Juventud.[58]

Reyes starts this essay by stating: "Let us leave aside Dr. Herrera's plasmogeny, which can also be called genoplasty. He was yet another wise scholar. By chance, while stirring his preparations, he found that he had created an amoeba, an elemental living cell, a unit of existence that managed to live several hours" ("Dejemos aparte la plasmogenia del doctor Herrera, que también podemos llamar genoplastia. Éste fue otro sabio. De casualidad, mientras revolvía sus preparaciones, se encontró con que había creado una amiba, una célula viva elemental, una unidad de existencia que logró vivir varias horas").[59] At the outset, Reyes refers to Herrera's plasmogeny—a science that the readers of *Contemporáneos* presumably were already acquainted with—in order to set the stage for a reflection on what life is and what it means to *create* it. Even though Reyes contends that he will *not* be talking about Herrera's plasmogeny, in what follows he does precisely this. The playfulness of this ironic start is accentuated by a word play (plasmogeny/genoplasty) that suggests an unforeseen connection: plasmogeny is somewhat related to a kind of cosmetic surgery that consists of modifying the chin to enhance facial features. Is the essay suggesting that plasmogeny is a mere cosmetic or superficial undertaking dealing with appearances? Are we led to think that Herrera's artificial beings are analogous to a cosmetically reconstructed face? Or should we simply take this word play as a slight, ironic derision one could expect from a young humanist intellectual referring to an older materialist scientist? At this point and throughout the essay, Reyes does not provide definitive answers and, on the contrary, prefers to leave the reader dwelling in the various ramifications of his assertions.

"Amiba artificial" goes on by stating:

> The news telegrams stress the accidental nature of the finding. And this nature, which seems to offend theology as little as possible, amounts to—if the news is accurate—the best guarantee. We admit that any mystery might be uncovered by chance. But on purpose, that is something else. A thousand hideous legions of prejudice whirl up! We are already entering the diabolical. . . . But here it was chance; let us take it calmly.

> *Los telegramas de la prensa insisten en el carácter casual del hallazgo. Y este carácter, que parece ofender a la teología lo menos posible, viene a ser también—cuando resultara exacta la noticia—la garantía mejor. Por casualidad, admitimos que se descubra cualquier misterio. Ya de propósito, es otra cosa. ¡Mil espantadizas legiones de prejuicios se levantan en torbellino! Entramos ya en lo diabólico. . . . Pero aquí fue una casualidad: tranquilicémonos.*[60]

Reyes alludes to the religious backlash provoked by plasmogenic experiments, which were construed as an attempt to demonstrate that God does not exist and did not originally create the world. Ultimately, religious "prejudices" deemed these experiments as "diabolic." Reyes suggests that there is a sense of paradox in such belief: what truly produces utter rejection is not the mere fact that life *can* be created in the laboratory, but the idea that scientists *intentionally* strive to do so, and therefore pretend to substitute for God. If a scientist would manufacture life by chance, then it would seem slightly more tolerable to religious people, because the Church's power and its ultimate sovereignty would remain intact. Plasmogenic experiments call into question religion not because they prove that life can be made, but because they insinuate that religion is no longer needed in the present world.

While Reyes satirizes the attitudes of institutionalized religions, he also ponders the fact that matter seems to self-organize according to underlying principles. Just as Herrera describes the surprising behavior of colpoids, Reyes talks about how the microscope reveals the "almost conscious" and "intelligent" demeanor of particles of matter: "If you have witnessed, under the microscope, certain reactions, you already know the almost conscious movement of the particles, as in a dance company where each dancer has to look for his partner, and has to follow a path to end up in a certain geometric position. Is this intelligent agitation not already like 'a tremor of life'?" ("Si habéis presenciado, al microscopio, ciertas reacciones, ya sabéis de ese correr casi consciente de las partículas, como en una danza de cuadrillas donde cada figurante tiene que buscar a su pareja, y cada uno atravesar un camino para ir a colocarse en un determinado lugar geométrico. Esta agitación inteligente ¿no es ya como 'un temblor de vida'?").[61] The minutest particles of matter already show the signs of what seems to be a principle of vitality that governs the organization of matter. Living beings might be in fact the product of one of those basic laws of matter: the law of crystallization. "What law is this," Reyes exclaims, "which imposes on matter a distribution already in conformity with the dogmas of decorative art, already in accordance with human feelings, in accordance with the rules of pleasantness and good taste? What can it be but a life principle, a first manifestation of the agitating and ordering mind?" ("¿Qué ley es ésta, que impone a la materia una distribución ya conforme con los dogmas del arte decorativo, conforme ya al sentimiento humano, conforme a las reglas de la simpatía y al buen gusto? ¿Qué ha de ser sino un principio de vida, una primera manifestación de la mente agitadora y ordenadora?").[62] More than conforming to arbitrary impulses, matter seems to naturally follow precise and harmonious vital principles, which even seem to be attuned to aesthetic or rational laws.

Accordingly, Reyes maintains that the Western intellectual tradition has always assumed the existence of some kind of first principle in order to understand being, even when philosophers have termed it variously throughout history: "Thinking about it, we already sensed it. Did we not know that in the beginning it was the Logos, the Word, the Breath, the Psyche, the Pneuma, the Thought, the Order or whatever you want to call it?" ("Pensándolo bien, lo presentíamos ¿No sabíamos que en el principio era el Logos, el Verbo, el Soplo, la Psique, el Neuma, el Pensamiento, el Orden o como se le quiera llamar?").[63] Thus, in consonance with the philosophy of his generational peer Antonio Caso, discussed in Chapter 2, Reyes subscribes to a sort of philosophical vitalism that postulates that life cannot be reduced to its material components without losing what essentially makes living beings *live*. Here Reyes was probably building on

Bergson's vitalism and his concept of *élan vital*, which had a decisive impact on Reyes's intellectual outlook and on the Ateneo de la Juventud as a whole.[64] But in any case, Bergson was only a recent representative of a long intellectual tradition that had repeatedly theorized an immaterial, a priori principle animating and fundamentally differentiating living beings from inorganic matter. For Reyes, as for other members of the Ateneo de la Juventud, the notion of a vital first principle afforded him the groundwork for conceptualizing how life separates itself from inorganic matter and eventually culminates in the higher faculties of the human spirit epitomized by literature.[65] As I have already shown, vitalist and humanist assumptions of this kind were staunchly contested by Herrera's philosophical and scientific endeavors.

However, paradoxically, at the same time that Reyes assumes the existence of an immaterial vital principle, he also suggests an idea that destabilizes his own humanist spiritualism and comes closer to Herrera's speculations. Alluding to Claude Bernard's proposal of the five basic characteristics of living beings (morphological unity, chemical unity, nutrition, reproduction, and specific form), Reyes contends that is not an easy task to successfully define life and ascribe it to an exclusive set of beings: "But it remains to be asked sincerely if life only exists in the so-called organic bodies. It is not easy to set limits. The five conditions (of Bernard) also occur in the mineral kingdom: atomic or electronic unity, nutrition, reproduction, specific shape of crystals" ("Pero falta preguntarse sinceramente si sólo existe la vida en los llamados cuerpos orgánicos. No es fácil limitar fronteras. Las cinco condiciones [de Bernard] se dan también en el reino mineral: unidad atómica o electrónica, nutrición, reproducción, forma específica de los cristales").[66] Here Reyes comes close to suggesting that "everything is alive," even crystals and other minerals, in accordance with Herrera's crucial proposition. Without ever referring to Herrera's philosophical thought—perhaps even ironically mocking him for being a crude materialist—Reyes seems to pose the same fundamental questions and propose similar reflections.

In addition to relativizing the biological distinction between life and nonlife, Reyes also alludes to the concept of universal life, which disrupts the traditional understanding of (biological) life even further. According to Reyes, the universe as a whole seems to display nutrition as one of the fundamental traits of living beings: "The formation of chemical bodies in the nebula and their radioactive decay are two phases of nutrition: anabolism and catabolism. Where does life begin and how does the scholar distinguish it from all the vibrant garbage seen through the microscope lens?"

("La formación de los cuerpos químicos en la nebulosa y su desintegración radioactiva son dos fases de nutrición: anabolismo y catabolismo ¿Dónde empieza la vida y en qué la distingue el sabio de toda la basura vibrante que ofrece el cristal del microscopio?").[67] In other words, the universe functions as a kind of organic body: it requires chemical reactions that first generate complex molecules (anabolism) and then breaks down those molecules in order to release energy available for use (catabolism). Is it justifiable, then, to claim that the universe is "alive" and shows the same sort of "almost conscious" and "intelligent" behavior that one can also see in matter particles under the microscope? Reyes does not go as far as Herrera in asserting this idea, but he is at least prepared to seriously entertain the thought.

Later in the essay, Reyes proposes a thought-provoking insight that relates plasmogeny to art: "Our scholar, then, does no more than suppress the 'likeness' of ancient rhetoric, thus giving us the 'created' metaphor, as today's poets call it" ("Nuestro sabio, pues, no hace más que suprimir el 'como' de la antigua retórica, y darnos así la metáfora 'creada', que dicen los poetas de hoy").[68] Reyes establishes a contrast between ancient rhetorical disciplines and avant-garde literary theory ("los poetas de hoy"), while simultaneously drawing a comparison between avant-garde poetry and plasmogenic experiments. But in what way can plasmogenists and other scientists concerned with manufacturing life in the laboratory be considered analogous to avant-garde poets who strive to create innovative metaphors ("metáfora creada")? What does it ultimately mean to compose a "metáfora creada"? Reyes did not fully develop his suggestions in this essay, but I will approach these questions by drawing from the contemporaneous aesthetic theorizations that he was probably referring to. Interestingly, as I will show, "Amiba artificial" suggests a conceptualization of poetry that seems to destabilize Reyes's own humanist understanding of culture and literature as the formulation of pure (human) experience.[69] If, as Gareth Williams suggests, Reyes strove to forge "a fully harmonious notion of humanistic culture over and against any inferior, abnormal, 'dark,' or potentially conflictive form of historical being," one could argue that the inorganic life underpinning art and human beings was one of those anarchic, conflictive forces that Reyes consistently tried to bracket off and neutralize in his writings.[70]

While the preference for creative and surprising metaphors is one of the main characteristics of avant-garde poetry and one that is shared by most, if not all, historical avant-garde movements, the Chilean poet Vicente Huidobro, founder of the Creationist poetic movement, published several manifestos in which he developed one of the most radical theorizations of what is ultimately at stake in crafting avant-garde metaphorical expressions. Huidobro argues that while traditional poetry strove to *imitate* nature by establishing explicit comparisons with the external world (what Reyes calls "the 'likeness' of ancient rhetoric"), avant-garde poetry transcends mimetic aspirations and endeavors to *create* a whole new world by acting as nature itself: "Man was never closer to Nature than now, now that he no longer seeks to imitate her in her appearances, but to do what she does, imitating her at the level of her constructive laws, in the achievement of a whole, in the mechanism of producing new forms" ("el hombre nunca estuvo más cerca de la Naturaleza que ahora que ahora que ya no busca imitarla en sus apariencias, sino hacer lo que mismo que ella, imitándola en el plano de sus leyes constructivas, en la realización de un todo, en el mecanismo de la producción de nuevas formas").[71] Huidobro contends that poets should aspire to reproduce nature's tendency to *create* new realities by continuous processes of differentiation and individuation. These poets would produce what Huidobro calls a "poema creado" composed of metaphors and images similarly "creadas"; that is, mimetically detached from any exterior object and aiming to create a new reality that stands by itself in the world.

By establishing a comparison between plasmogeny and avant-garde poetry, Reyes suggests that Herrera's artificial beings are a sort of vanguardist poem in the sense that both are "created" realities, apparently detached from external reality. Plasmogenists and avant-garde poets "do no more than suppress the 'likeness' of ancient rhetoric," namely, they do not aspire to imitate the products of nature but to attune to its productive potentiality and put it to work to create new realities (poems or artificial beings). In the same way that avant-garde poets do not strive to "sing of roses" but aspire to "let them flower in the poems," as Huidobro puts it, plasmogenists are not interested exclusively in formulating an abstract theory but also in creating a material embodiment (colpoids, for example). This means that art and plasmogeny are processes of engagement with nature's forces that are not primarily guided by the objective of mimesis/representation but can in fact be conceived of as *material* interactions with the cosmos that ultimately create new bodily experiences. In other words, both Herrera and poets such as Huidobro are determined to connect with the universal flows of inorganic life in ways that allow for the creation of new perceptions (visual, auditory, and so forth) and cosmic interactions. Manufacturing a colpoid in the laboratory or crafting a creative metaphor are both ways of altering the usual paths of universal life in order to make room for innovative arrangements of matter and force.

Perhaps plasmogeny and poetry are best understood as techniques for habilitating *becomings*: "To become is not to attain a form (identification, imitation, Mimesis) but to find the zone of proximity, indiscernibility, or undifferentiation where one can no longer be distinguished from *a* woman, *an* animal, or *a* molecule—neither imprecise nor general, but unforeseen and non-preexistent, singularized out of a population rather than determined in a form."[72] Poetry finds a zone of undifferentiation within language by creating a kind of "foreign" or artificial language made up of neologisms, creative metaphor and images, innovative syntax, and so on. In some cases (think of the last section of Huidobro's *Altazor*, for example), avant-garde poetry may end up producing an "unforeseen and non-preexistent" language, a completely invented language that has no external referent: nothing more than rhythmic sound patterns that resemble nonhuman expression. There is here a becoming-animal of the poet who sheds his proper human form and, as Deleuze puts is, "can no longer be distinguished from *an* animal." Similarly, as already explained, plasmogeny reaches a space of indiscernibility between the living and the nonliving, the organic and the inorganic: looking at Herrera's experiments it becomes impossible to differentiate between living stuff and *a* molecule, between humanity and other kinds of beings. In brief, evoking new materialist theoretician Elizabeth Grosz's approach to art, perhaps one could say that plasmogeny and poetry are ways of engaging with and intensifying forces that are not strictly *human*: these are nonhuman forces (animal, molecular, cosmological, and so forth) that decenter the traditional understanding of the human and literature in Western thought, including in Reyes's own work.[73]

Reyes's essay insinuates that plasmogenic experiments are a creative or artistic—not primarily scientific—endeavor. The arts and sciences are ways of engaging the world that ultimately have different approaches: while science tends to extract variables and constants that allow us to predict and measure, the arts arrange sensations and affects that proliferate beyond expected limits.[74] When one encounters a colpoid or an avant-garde poem, one encounters a block of sensory signals that is first and foremost concerned with facilitating an unusual sensory experience. Herrera seemed to suggest that scientific theories and philosophical concepts were ultimately incapable of *explaining* or reducing plasmogenic experiments to a set of stable propositions. He often emphasized the importance of experiencing his experiments *in person*, and even volunteered to send samples to anyone who might be interested. For Herrera, anybody who encountered his experiments would undoubtfully undergo an unprecedented sensory experience and would generate a kind of *bodily* understanding of abstract

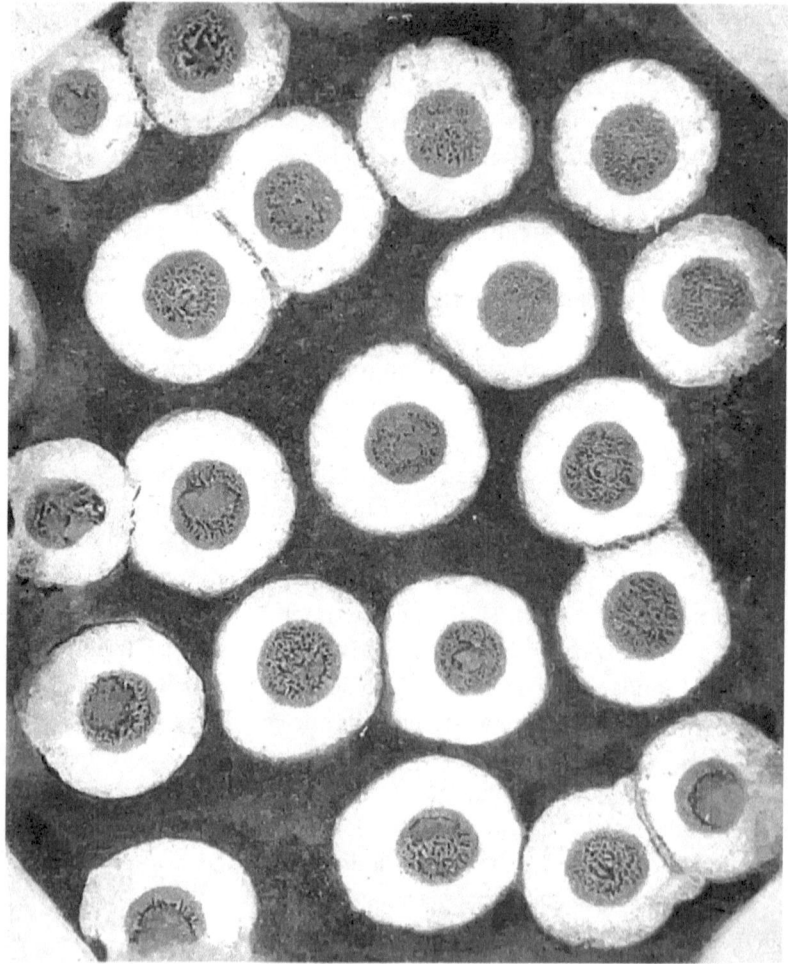

FIGURE 1.4. "Drops of alcohol on alkali silicate, and ultramarine blue," from Alfonso L. Herrera, *Una nueva ciencia: La plasmogenia*, 1926

concepts. His writings were accordingly full of descriptions, images, and photographs that attempted to convey the surprising colors, movements, odors, and textures acquired by the artificial beings fabricated in his laboratory.

Reyes's brilliant insight compels us to conceive plasmogeny as an artform. It is not only that, as Herrera proposed, plasmogenic experiments could have consequences for every branch of human knowledge, but also that these experiments could themselves be considered a kind of avant-garde art. Even though Herrera himself never deemed his experiments to be art, there is no doubt he appreciated their aesthetic appeal and the creative possibilities they open for human perception and thought. It is not farfetched to consider plasmogenic experiments as bioart: an art practice that does not restrict itself to imitating thematically living organisms but works instead with life's processes as the basic material of art.[75] As art historians have argued, bioart first emerged in the 1930s—around the time Herrera was busy performing experiments and writing poems—when photographer Edward Steichen presented an exhibition of genetically altered plants in the Museum of Modern Art.[76] This project was part of the avant-garde effort of expanding the notion of art by including non-Western traditions, everyday objects, automatic writing, "metáforas creadas," and other artistic strategies.

As Mitchell argues, bioart strives to generate a specific kind of reaction: "the gallerygoer's sense of becoming-a-medium—the sense, that is, of being part of a biological milieu that has logics of transformations that exceed the gallerygoer's own goals and interests."[77] Bioart does not aspire to provoke ideas or representations in the spectator's mind, but to generate a bodily sense of the uncanny: being at the same time familiar and deeply unfamiliar with life's processes, acknowledging oneself as part of life but also recognizing the inability to predict or control its becomings. If biotechnology produces the fiction that vital processes are ultimately controllable and optimizable, bioart endeavors to challenge that idea by suggesting that life ends up overflowing every technique that seeks to restrain it. As already mentioned, what Herrera hoped his experiments would accomplish is analogous to bioart's approach: interacting with these "intelligent" and "conscious" particles, the spectator of Herrera's experiments would challenge the ontological and biological privilege attributed to humanity and would recognize themselves as part of the universal flows of inorganic life that vastly exceed human beings.

Thus, plasmogeny and poetry might be seen as spaces that propitiate the prolongation of the inorganic life that has historically been "determined in a form," that is, confined to living organisms. Both are activities that reveal not only the inorganic foundation of all organic matter and the very impossibility of making an absolute distinction between organic/inorganic, but also that life cannot be reduced to the kingdom of living organisms or to any other specific area. Furthermore, this conceptualization of plasmogeny and poetry undermines the humanist understanding of culture and literature posed by Reyes and many other artists, which assumes that culture is a practice exclusively concerned with "higher," spiritual faculties that must bracket off "lower" material beings and processes. As we will see next, Herrera's own poetry compiled in *Murmullos del universo* is an exploration that, while thematically portraying the notion of inorganic and universal life, remains largely grounded in humanist, traditional frameworks.

MURMULLOS DEL UNIVERSO

Murmullos del universo was published posthumously in 1982 by Herrera's youngest daughter, María Amalia Herrera y Estrada. According to her account, Herrera wrote this book of poems during the last decades of his life and even prepared a typed copy with the aim of publishing it. What finally appeared forty years later is a facsimile edition of said typed copy, which also contains many images and photographs selected by Herrera himself. *Murmullos del universo* is composed of five parts (cantos) devoted to a wide array of subjects such as nature, space, personal relationships, and social issues, among others. Comprising more than half of the work, the first two cantos ("Canto I: En el infinito" and "Canto II: Murmullos en las estrellas gigantes") are a sustained reflection on natural phenomena including the sea, the jungle, volcanoes, trees, animals (including poems dedicated to toads and walruses), springtime, the stars, the Milky Way, and so on. The third canto ("Canto III: Murmullos estelares de amor") is devoted to romantic and fatherly love and includes poems about Herrera's wife, daughters, and grandchildren. While the fourth canto ("Canto IV: Rumores internos del pensamiento") is an interlude concerned with the implications of the act of thinking itself, the last canto ("Canto V: El llanto y el dolor en la infinita sombra") contains an exploration of political and historical issues such as the Mexican nation, World War II, the Pyramid of the Sun in Teotihuacan, and the modern metropolis.

Insofar as it preserves a traditional approach in terms of rhyme and meter, *Murmullos del universo* is not a work of formal experimentation akin to the avant-garde aesthetics and cultural renewal pervasive in post-revolutionary Mexico. Herrera belongs to the generation of *modernismo literario* both in terms of the time of his birth and his aesthetic outlook: in general terms, his poetry shares more formal and thematic traits with the modernistas—Enrique González Martínez, Amado Nervo, and especially Manuel José Othón—than with the innovative or avant-garde groups Contemporáneos and Estridentistas. Even if *Murmullos del universo* does not generally display avant-garde techniques and objectives ("metáforas creadas" and other strategies), it does introduce complex poetic structures that end up defamiliarizing our perception of the world. What stands out in particular is how Herrera's poetry incorporates scientific concepts and terminology in order to create a new poetic language that is unheard of in the postrevolutionary context. It could be said, then, that *Murmullos del universo* establishes a zone of undifferentiation between scientific and poetic discourses in order to create new ways of engaging with the universe.

In his prologue to the work, Herrera points out what is at stake for him as a scientist in the endeavor of creating poetry. He argues that science and poetry are complementary undertakings in the process of comprehending the ultimate foundations of the universe: "unexpected flight from peak to peak, from mountainside to mountainside, with the spread wings of science and poetry holding hands" ("inaudito vuelo de cumbre en cumbre, de vertientes en vertientes, con las desplegadas alas de la ciencia y la poesía tomadas de las manos").[78] Poetry, particularly a kind of *scientific* poetry, can be considered a natural product of science thoroughly exploring everything that exists. In this way, science and poetry strive together to provide a glimpse of the fabric of the whole universe, its origin and ending, its minutest and grandest features, its diverse phenomena: "Let us unite science, philosophy and poetry, following Renan's advice, to form a humble assemblage of truth and harmony, of research and art" ("Unamos la ciencia, la filosofía y la poesía, siguiendo el consejo de Renán, para formar un humilde conjunto de verdad y armonía, de investigación y arte").[79] The scientific study of any part of the universe leads directly to pondering the unity and interconnection between all of its parts through processes of transformation and causality. In sum, science and poetry both reveal and explore the becomings of universal life.

Herrera devotes several poems in *Murmullos del universo* to singing the praises of science and its inextricable connection with poetry. One example is the poem titled "A los maestros," which is an invitation extended to Mexican teachers to embrace scientific thought. This poem combines a traditional form (stanzas of four eleven-syllable lines with ABAB consonant rhyme) and conventional imagery to make an apology for science. Employing rationalist symbolism, scientific knowledge is equated to a sun that illuminates every corner of the universe and dissipates doubt and fanaticism. But science is not conceived of as a positivist endeavor that is detached from a humanistic approach to the world, as was traditionally conceptualized by the positivist intellectual elite during the Porfiriato. On the contrary, "A los maestros" emphasizes the interconnectedness of poetry and science: the result of scientific practices is in fact an "unlimited poem" that embraces and *poeticizes* "The unity of the heavens and the earth / The total evolution of the Universe!" ("La unidad de los cielos y la tierra / La evolución total del Universo!")[80] All universal phenomena and its constant transformation should be the focus of the joint action of science and poetry: "From the infusorium to man and microplasma, / From the atom to the granite mountains, / From the colloidal world of protoplasm, / To the world that bursts into infinity!" ("Del infusorio al hombre y microplasma, / Del átomo a los montes de granito, / Del mundo coloidal de protoplasma, / Al mundo que revienta en lo infinito!")[81]

Herrera not only devotes poems to science and incorporates scientific terminology, but also occasionally cites scientific works to support his poems. For example, at the end of "La música universal" there is a footnote that clarifies that, in line with recent investigations by Spanish astronomer Comas Solá and reinforcing the Pythagorean theory known as the Harmony of the Spheres, celestial bodies have been found to emit music; what is more, according to modern physiology the human brain generates ideas in the form of chords or harmonics. This seems to be the scientific evidence supporting the auditive references pervasive in "La música universal" and throughout Herrera's poetry. In fact, it could be argued that, as its name suggests, *Murmullos del universo* wagers for recreating the experience of *listening* to the universal life's music. Ultimately, according to Herrera, the poem must be "a pale glimmer, a dying echo of universal life" ("un pálido destello, un eco moribundo de la universal vida").[82] Poetry is conceived of as a rhetorical device that reproduces and magnifies the music, the echoes, murmurs and voices of the universe.[83] For example, "La música universal" starts by imploring that the sounds of the universe "sing" or "murmur" in the poet's mind. The "voice of the universe" is compared to a kind of "music," and "life" is presented as the "monumental throat" that produces said universal music. At the core of "each verse," there is a "romantic stanza" that resists or even daunts the infinite magnitude of the universe.[84]

The poem "Supremas revelaciones de los sonidos" also begins by extending an invitation to listen to the universe: "Listen to the sounds and rumors / Without origin or end, of the burning focus, / Of the atom and the parent ions; / The deep vibration of the torrent, / The thunderous cries of Pluto, / The notes of life, the indescribable / Eternal movement in the Infinite!" ("Escuchad los sonidos y rumores / Sin origen ni fin, del foco ardiente, / Del átomo y los iones genitores; / La vibración profunda del torrente, / Los gritos de Plutón atronadores, / Las notas de la vida, el indescrito / Eterno movimiento en lo Infinito!")[85] But later on, the reader realizes that the poem is actually representing the experience of listening to classical musicians, particularly of the romantic era: "These are the notes that Schubert chains together" ("Son las notas que Schubert encadena").[86] By doing so, the poem depicts how romantic music conflates sonically with the sounds and murmurs of the universe, as if both were a product of the same cosmic forces. This is why classical music—not avant-garde art—provides a template for Herrera's own poetry and its aim of condensing the melodies of universal life. As he states in the prologue, "Music will lend me its powerful wings, its trills and its tenderness. A stanza will wander through the poem, it will be repeated, with variations, like a musical theme, as in the fugues of Bach and Beethoven" ("La música me prestará sus potentes alas, su gorjeo y su ternura. Una estrofa vagará por el poema, será repetida, con variaciones, como el tema musical, como en las fugas de Bach y de Beethoven").[87] Herrera's reliance on strict rhyme and metric schemes is thus a crucial strategy for reproducing acoustically the melodies of the universe, from the microscopic world of atoms to the macroscopic sphere of planets and celestial bodies.

Poetry allows Herrera to recreate the universal life existing both in the innermost spheres of matter and the farthest regions of the universe. In this vein, Herrera devotes a poem to the Andromeda galaxy, which is considered one of the farthest objects one can see with the naked eye. In this galaxy the clear manifestation of universal life is already there: "In the tenth century discovered / By the Persian Sufi, you are not dead! / Inside your womb lives burning / The enormous whirlwind of atoms" ("En el décimo siglo descubierta / Por el persa Sufí, tú no estás muerta! / Adentro de tu entraña vive ardiente / El torbellino de átomos ingente").[88] Here, the murmurs of the universe—visually represented as "pálidos destellos" coming from a galaxy located millions of light years from Earth—reveal the lively movement of atoms that is found throughout the universe. Similarly, the poem "El nacimiento de una estrella" represents outer space as dark, vast, and even dreary: "How sad is the silence of the abstract! / How sad its darkness, eternal cold, / The loneliness of the cosmos, the abode / Of Harpocrates, the mute and shadowy god!" ("Qué triste es el silencio de lo abstracto! / Qué triste su tiniebla, eterno frío, / La soledad del cosmos, la morada / De Harpócrates, el dios mudo y sombrío!")[89] But this same place is the very cradle of life: "But be aware, brothers in life, / shadows of shadows, that in the frozen night / the parent plasma, hidden sap, / like an electron it stirs in nothingness" ("Pero sabed, hermanos en la vida, / sombras de sombras, que en la noche helada / el plasma genitor, savia escondida, / como electrón agítase en la nada").[90] Herrera once again refers to the idea that the flow of universal life unifies every single being in the cosmos as "brothers in life."

In many instances, *Murmullos del universo* also directs our attention to "infinite microcosmic life" ("la vida microcósmica infinita"): the microscopic composition of nature and the whole universe. The poet recognizes that humanity is constituted and traversed by this endless microscopic life: "We are only ashes of the worlds, / Of atomic colonies reflections, / That form our body, of deep / Swarms of ions, the phalanx / Of the fertile protein cycles" ("Sólo cenizas somos de los mundos, / De colonias atómicas reflejos, / Que forman nuestro cuerpo, de profundos / Enjambres de los iones, los artejos / De los ciclos proteínicos fecundos").[91] These lines contain terms that suggest the small scale of this world ("atómicas," "iones," "proteínicos"), as well as words that accentuate the notion of the multiplicity and interconnectedness of universal life ("colonias," "enjambres," "artejos"). The interrelation of all these minutest particles is also highlighted in these lines by the use of the rhetorical figure of enjambment, that is, the extension of a syntactic structure beyond the break of a line. This rhetorical figure creates an effect of interconnectedness by forcing the reader to unify two different lines, while also producing a sense of trepidation and confusion when the reader is required to keep reading line after line in order to comprehend the overall meaning of the poem.

In a poem dedicated to his granddaughter, Herrera starts by proclaiming the love and tenderness inspired by his "little granddaughter" ("nietecita"): "It weighs like a world on me; I love her so much / That I sleep her sweetly in my arms. / Shy and sweet, rosy and beautiful / Lights up my old age like a star ("Me pesa como un mundo; tanto la amo / Que la duermo en mis brazos dulcemente. / Tímida y dulce, sonrosada y bella / Alumbra mi vejez como una estrella").[92] But immediately after this romantic expression of affection the poem unexpectedly zooms in to contemplate her chemical composition: "And her matter is deep chemistry, / Cycle of polypeptides, substance, / Electronic whirlwind floods her, / Hidden Nile that is born from the ether / And becomes life, cries and screams" ("Y su materia es química profunda, / Ciclo de polipéptidos, sustancia, / Torbellino electrónico la inunda, / Oculto Nilo que del éter nace / Y en vida se convierte, llora y grita").[93] The little girl's body consists of the universal matter that arises from the ether—the protoplasm of the universe, according to Herrera—and transforms into a human being that screams and cries. In this way, the poem achieves a disruption of the reader's expectations: first, the reader anticipates a rather straightforward poem of a grandparent's love deploying conventional terms ("la amo," "dulcemente," and so on), but as soon as the reader is invited to venture into the microscopic composition of the human body the poetic language becomes rarified by incorporating scientific terms ("polipéptido," "electrónico," "éter"). This alteration of the reader's expectations does not create a sense of disavowing the affection displayed earlier in the poem, as if suggesting that parental love does not materially exist or is only a sentimental fiction. On the contrary, the poem seems to imply that parental love is even more intense because it is ultimately grounded on the chemical and physical realities of universal life. Loving your grandchildren, Herrera suggests, is one of the most profound ways of admiring the endless manifestations of the inorganic life in the cosmos.

Murmullos del universo not only penetrates the microscopic and macroscopic worlds, it also aspires to recreate past iterations of the universe and even its future possibilities, in consonance with Herrera's own utopian projects. In "La verdad es el cáliz del infinito," the poetic "I" addresses the Earth, seeking to learn its past: "Let me dig into your past, Earth, / Let me penetrate the great gorge / Where roaming they go to the unknown / Those who are, those who come, those who were, / The kings, the slaves, the Templars, / The geniuses, the impostors, the damned, / The Borgias, the Descartes, the Giordano, / The Mammoths, the Sequoia, the bacteria!" ("Deja que escarbe en tu pasado, Tierra, / Que penetre en el gran desfiladero / Donde rondando van hasta lo ignoto / Los que son, los que vienen, los que fueron, / Los reyes, los esclavos, los Templarios, / Los genios, los tartufos, los malditos, / Los Borgia, los Descartes, los Giordano, / Los Mamuts, las Sequoia, las bacterias!")[94] The Earth's crust contains matter from the decomposed bodies of extinct species such as the mammoth, and of every human being that has ever existed, notwithstanding their social status or accomplishments during their lifetime. Moreover, this same matter will recompose and take new forms in the future: "those to come" ("los que vienen"). The matter of the universe is immortal in this precise sense: it transforms and takes different shapes, but it never passes away definitively: "If everything is immortal we never die / And when we wake up from deep sleep / Universal Life, our apotheosis, / Smiling awaits us among the tombs!" ("Si todo es inmortal nunca morimos / Y al despertarnos de profundo sueño / La Vida Universal, nuestro apoteosis, / Sonriendo nos espera entre las tumbas!")[95] As we will see in Chapter 2, the fact that universal life is endless and everlasting allows Herrera to imagine future resurrection of the past.

Even though *Murmullos del universo* constitutes a *thematic* engagement with the cosmic flows of inorganic life—the prolongation of the songs and echoes of universal life that undermine human privileges—it does so relying mostly on *formal* strategies that underpin the humanist conception of poetry. In Western societies, art has been historically conceptualized as the quintessential human faculty, a spiritual endeavor that distinguishes not only artists within society, but also humanity from other living beings and the rest of nature. During the first decades of the twentieth century, while Herrera was at work creating his poetry, the historical avant-garde movements endeavored to put pressure on these long-standing humanist assumptions, critiquing the social institution of art—based on its alleged "autonomy" from other social spheres—and calling for art's complete immersion in society and life at large.[96] This call was accompanied by a renewal of artistic forms and approaches that unmistakably signaled a radical break with tradition and the past. For instance, as Reyes suggests in "Amiba artificial", Vicente Huidobro and other poets strove to *create* poetic realities that would imitate the functionality of nature, not its products. Similarly, as Rubén Gallo has demonstrated, a handful of avant-garde writers in Mexico aimed at disrupting traditional artistic media by engaging with the nonhuman force of technology (cameras, typewriters, radio, cement, and so on).[97]

Even though Herrera's poetry is innovative in its incorporation of scientific terminology and notions, it does not build on avant-garde approaches or incorporate leading-edge artistic techniques. On the contrary, it mostly remains attached to the kind of humanist, romantic framework and traditional formal strategies endorsed by *modernista* poetry. This seems to contradict Herrera's purpose of poetically reproducing the modes of expression of universal life—the songs, murmurs, and echoes articulated by every being in the microscopic sphere of particles and the macroscopic world of galaxies, including mineral, vegetal, and animal beings. Such an intention aims at destabilizing the idea of humanity—and, particularly, the arts—as the privileged, unique source of expression in the universe. In this way, *Murmullos del universo* shows a structural, unresolved tension between its traditional, humanist formal composition and its radically materialist, non-anthropocentric outlook, which is of course the same tension that lies at the center of Herrera's intellectual project and biocosmism at large.

CONCLUSION

Plasmogeny, as a discourse that examines universal life, aspired to have a crucial impact on a wide diversity of spheres: from practical applications (technology, medicine, and so on) and scientific knowledge to philosophical discourse and artistic endeavors. Herrera produced a multilayered discourse that encompassed diverse projects: it entailed, at the same time, a scientific understanding of the origin of life, a project of biopolitical transformation of humanity, a materialist philosophy, and a poetic recreation of the cosmos. As I have argued, some of these facets fundamentally contradicted each other: on the one hand, the project of manufacturing a new race of "supermen" in the laboratory was founded on the biopolitical purpose of hierarchically distributing desirable forms and residual forms. This utopian project envisioned a perfect westernized White civilization that exerts control over other races, living beings, and the whole cosmos. On the other hand, according to Herrera, his plasmogenic experiments suggested "a chain of union and causality" that connected universal matter in continuing processes of transformation. This philosophical notion of universal life destabilizes hegemonic differentiations between life and nonlife, the organic and the inorganic, humanity and other living beings.

I have argued that Herrera's work was an attempt at addressing the various problems and questions raised by his plasmogenic experiments through the manifold approaches of science (the theory of the origin of life), philosophy (the notion of universal life) and the arts (his poetry collection *Murmullos del universo*). Herrera conceived all these endeavors as different forms of engaging with the inorganic life of the cosmos. He was concerned with understanding and interacting with the inhuman, cosmological forces that manifest everywhere from the molecular level to outer space. This becoming-cosmic of humanity aspired to open up a utopian line of flight away from the hegemonic categories of Western culture. The next chapter will continue the examination of Herrera's work by focusing on his utopian projects for resurrecting past forms of life and his theorizations concerning the "chemical ethics" of universal life.

CHAPTER 2

Resurrecting the Past

Animality and Chemical Ethics
in Alfonso L. Herrera

In 1928, Alfonso L. Herrera requested from the Mexican ambassador in the United States a copy of the dinosaur fossil Diplodocus Carnegiei, named after the American businessman Andrew Carnegie, who had funded this fossil search in 1899. Carnegie's Diplodocus became one of the first and largest complete dinosaur fossils to be displayed as a centerpiece in museums with the aim of attracting the attention of mass audiences. In the next three decades after the discovery, Carnegie and his wife donated copies of this huge fossil to a dozen European and Latin American museums, including the Museo Nacional de Historia Natural run by Herrera, which solidified its fame as the dinosaur fossil most observed around the world.[1] At that time, no complete dinosaur fossils had been found in Mexico—much less one of the dimensions of Diplodocus—and they had not been exhibited in museums. According to his testimony, once Herrera learned of the possibility of obtaining a copy of Diplodocus, he assumed that it was a bronze copy that he might display outdoors in the Chapultepec botanical garden.[2] His plan was to place the fossil near a lake in the forest, so as to simulate the actual habitat and vegetarian diet of this immense reptile. However, the copy that was finally obtained in 1930 was made of plaster and was installed in the Museo Nacional de Historia Natural, to the amazement of visitors who, for the first time, laid eyes on a fossil with these impressive features.[3]

In the early twentieth century, exhibitions featuring dinosaur fossils contributed to the constructed illusion that the extinct past could be "revived" so that viewers could see and admire it first-hand. It was this sense of proximity and contact with the distant, utterly different past that attracted and continues to attract thousands of people to exhibits of gigantic dinosaur fossils in museums around the world. Herrera's original plan—placing the Diplodocus fossil outdoors with the aim of imitating its actual modus vivendi—was a curatorial proposal that tended to further enhance the illusion of resuscitating the past and displaying it in the present world. Walking through the Chapultepec forest and suddenly observing the appearance of the enormous figure of the Diplodocus near a lake would be an impressive way for the masses to learn the biological traits of this reptile and, more generally, to understand the drastic transformations and the evolutionary changes that life on Earth has undergone. Herrera's project was in accordance with his speculations analyzed in the first chapter about the need for biological museums to surpass the taxonomical classification of living beings, which only immobilizes them as "dead things" that belong to the past. Rather, museums of natural history needed to exhibit and "bring to life" the dynamic principles that have underpinned nature and biological life from its origin to the present. The plan to "revive" the Diplodocus in Chapultepec was an expression of Herrera's point of view that museum exhibits should have a massive and transformative impact on public awareness in order to ensure their usefulness.

The intention of bringing the past back to life guided Herrera's initiatives beyond obtaining the Diplodocus fossil and his involvement with the museum institution. In a way, Herrera's utopian project of biologically creating a perfected humanity in the laboratory—discussed in Chapter 1—was complemented by his plans to resurrect extinct forms of life. To that end, he conceived two different plans: first, he undertook the task of recreating the immediate biological antecessor of humanity through the hybridization of humans and apes. This experiment would allegedly prove Darwin's evolution theory and have transcendental implications for human civilization. The second project built on the idea that, since the matter populating the universe is neither created nor destroyed, there must still exist somewhere in the cosmos the structuring particles of past biological species and even individual human beings. Herrera imagined that future technological devices would be able to gather those particles to reconstruct bygone forms of life. Both projects are founded on the assumption that humanity has developed the ability and knowledge to control and manipulate nature's temporality, understood as a teleological and necessary trajectory: human beings will not only synthetically create future living forms in order to accelerate nature's progression, they will also retrace nature's steps and recreate past forms of life. The fact that, according to Herrera, there is no essential difference between humans and other living beings, between the organic and the inorganic, brings up the possibility of technologically rearranging the universal flow of matter for the purpose of "improving" nature.

This chapter discusses the philosophical and political implications of Herrera's plans for resurrecting extinct forms of life. The first part focuses on the scientific and political reasons that motivated Herrera to attempt the hybridization of humans and apes by artificial insemination. Inspired by a teleological approach to evolution, the purpose of proving the biological proximity between humans and primates ended up reifying animality and introducing a racial hierarchy *within* the human species. However, I argue that Herrera's notion of chemical ethics, developed in *Filosofía etérea* (1919) and elsewhere, formulates a radically materialist form of vitalism that undermines his own teleological and mechanistic conceptualization of life and evolution. Drawing from new materialist approaches to the agency of materiality, I examine how the notion of chemical ethics further elaborates on the contingency and vital agency of inorganic matter across the universe. I then contrast Herrera's approach to the natural world with Antonio Caso's vitalist philosophy and Fernando Ocaranza's biological thought, with the aim of assessing Herrera's position in postrevolutionary Mexico. Finally, the chapter explores how Herrera's intellectual project relates to Diego Rivera's biocosmist visual work.

HYBRIDIZATION EXPERIMENT

In 1933, Herrera published in Spain a pamphlet entitled *El híbrido del hombre y el mono*, which details his efforts to produce human beings' now extinct "direct ancestor" by artificially inseminating a female chimpanzee with human semen. Herrera begins the pamphlet by setting out the scientific basis that makes this experiment possible. The production of individuals (animals or plants) as a result of the mixture of two different species is a subject that has been studied since the seventeenth century by scientists such as Gmelin, Linnaeus, Kolreuter, and Darwin, among others. Herrera observed that hybrids had been obtained from a diversity of related species, including the mixture of turnip and radish, starfish and sea urchin, horse and donkey, lion and tiger, which suggests that "there is unlimited number of possible hybrids" ("hay ilimitado número de híbridos posibles").[4] The technique of artificial insemination, first discovered by Spallanzani and Hunter in the eighteenth century, has already proven successful in both cattle breeding and human fertilization and promises to be a viable method in the hybridization of the monkey and the human being. Well aware of the incredulity that this experiment might cause, Herrera affirms that the aforementioned scientific background should "lead the reader to believe that there is nothing absurd in our project" ("llevar al ánimo del lector el convencimiento de que nada tiene de absurdo nuestro proyecto").[5]

Herrera's pamphlet continues by describing how he became involved in the recent efforts to produce the biological mixture of the monkey and the human being. Herrera first mentioned the possibility of this hybrid in his book *Zoología* (1924) and later, according to his testimony, he took on the task of proposing this experiment to zoological parks in the United States, "but they are afraid of attacks from the press and of fanaticism, and my project has been abandoned" ("pero se tiene miedo de los ataques de la prensa y del fanatismo, y mi proyecto se ha abandonado").[6] During those same years, Herrera learned of the hybridization experiments carried out by doctor Ilya Ivanov (1870–1932) with funding from the Soviet government. Despite the secrecy with which these experiments were conducted, the news of this unusual project reached numerous newspapers around the world, in some cases mistakenly reporting that the hybrid of the monkey and the human being had been achieved.

Ivanov was a highly esteemed doctor who had perfected and popularized the artificial insemination of cattle and horses in late nineteenth-century Russia. In 1925, during the time of utopian effervescence that followed the 1917 Revolution, Ivanov managed to convince the Bolshevik authorities to sponsor his plans of biologically mixing humans and chimpanzees. The project also found support from the Pasteur Institute of France, which granted access to its research center located in Kindia, French Guinea, where Ivanov had access to plenty of monkey specimens. Ivanov first tried to impregnate African women without their consent using chimpanzee semen, but his plans were met with resistance from the French authorities. Later, Ivanov decided that the ideal method was to artificially inseminate three female chimpanzees with human semen, which proved unsuccessful as well. Ivanov later took chimpanzee specimens to the subtropical region of the Soviet Union to continue his experiments at a newly created primate study center, but most of the animals died on the journey or failed to adapt to their new habitat.[7]

Herrera became aware of this experiment through the news media and he subsequently came in contact in 1927 with one of Ivanov's main defenders, the American lawyer, biology enthusiast, and atheism promoter Howell S. England (1864–1938), who set out to raise funds to sponsor the continuation of the Soviet doctor's plans. Herrera not only offered his support to England's efforts, but also attempted to carry out the hybridization experiment on his own. In his 1935 article "Mi labor revolucionaria en la enseñanza,"

Herrera provides a summary of his initiatives concerning the teaching of secular and scientific principles through publications, the academy, as well as museums, zoos, and botanical gardens. Herrera attaches great importance to the fact that he requested a chimpanzee for the Chapultepec forest zoo, a biological species that "had never been seen before in Mexico and it was the delight of the curious visitors, showing its great intelligence and eating at a table, with cutlery, like a child" ("nunca se había visto antes en México y hacía la delicia de los curiosos, demostrando su gran inteligencia y comiendo en una mesita, con cubiertos, como un niño").[8] He later writes, "I ordered another specimen, a female, to try to produce a hybrid with the man, by Dr. Ivanoff's procedure, which was not done in the end" ("encargué otro ejemplar, hembra, para intentar la producción de un híbrido con el hombre, por el procedimiento del doctor Ivanoff, lo que no se hizo al fin").[9] According to him, Mexico had the ideal climatic conditions to keep the chimpanzees alive and perform the artificial insemination. Herrera attributes his failure to the "lack of material support" ("falta de apoyo material") and the fact that the Directorate of Biological Studies finally disappeared in 1929.[10]

Significantly, the reasons that justified this radical biological intervention were slightly modified depending on the scientific, cultural, and political context in which it was inscribed and discussed, taking on new implications, promises, or uncertainties.[11] In the Soviet context, Ivanov's project was supported by Bolshevik politicians and scientists seeking a drastic transformation of the socio-economic and political system and even of human biology and psychology. Although Ivanov limited himself to proclaiming the scientific feasibility of the experiment, the Soviet authorities quickly gleaned that the project's success would function as an absolute refutation of religion's teachings regarding the origin and evolution of the human being.[12] Obtaining a hybrid of the monkey and the human being would not only provide irrefutable arguments for the Darwinian theory of evolution and for materialist positions in general, it would also act as propaganda against the political power of religious institutions and in favor of Soviet revolutionary initiatives. The success of the experiment would mean, in short, a living proof that hierarchies considered natural or essential needed to be overthrown to open up space for the radical renewal of society led by the Bolshevik government.

Undoubtedly, the justification provided by the Bolshevik authorities was equally relevant in the American and Mexican context. England was a proselytizer for Darwinism and member of an association devoted to atheist propaganda (Association for the Advancement of Atheism), so his interest in Ivanov's experiment was based on considerations akin to those propounded in the Soviet Union. Similarly, Herrera placed great hopes in this experiment as further proof of evolutionary doctrine and of his theory of the inorganic origin of life. The hybridization of monkey and human would undeniably illustrate the natural path of evolution from the inorganic, passing through the first cell, until reaching homo sapiens. As Herrera points out in the initial pages of his pamphlet: "It is not true, readers and brothers, that God yanked us out of nowhere: we come from the underworld of minerals; then plants, and finally animals, as all biology proves and as it is shown with facts, for example, in the great Museum of New York" ("No es cierto, lectores y hermanos míos, que un Dios nos haya extraído a tirones de la nada: venimos de los bajos fondos de los minerales; después, de los vegetales, y al fin, de los animales, como lo prueba toda la biología y demuestra con hechos, por ejemplo, el gran Museo de Nueva York").[13]

Likewise, Herrera's pamphlet underscores the fact that atheist and anticlericalist ideas constitute a crucial motivation for this experiment's desirability. Herrera includes a section entitled "Lamentable and criminal influence of the church" ("Lamentable y criminal influencia de la iglesia") that begins with the following statement all in capital letters: "The church is opposed with all its might to Darwin's theory, to plasmogeny, to the crossing of man and the monkey, to contraceptive means and, in general, to all progress in science that is highly beneficial to humanity and that at the same time attacks biblical dogma" ("La iglesia se opone con todas sus fuerzas a la teoría de Darwin, a la plasmogenia, a la cruza del hombre y el mono, a los medios anticoncepcionales y, en general, a todo progreso de la ciencia que sea altamente benéfico para la humanidad y que a la vez ataque el dogma bíblico").[14] Certainly, Herrera's anticlericalism echoed a predominant stance both in the postrevolutionary elites' enlightened discourse and in popular sectors' skeptical or satirical gestures.[15] The two great tendencies of revolutionary anticlericalism—a mainstream and diverse stream of moderates or reformists, and a minority trend of radicals or iconoclasts—shared the rejection of the Church and Catholic dogmas, but had notable differences in their ultimate goals and the means of achieving them.[16] While the moderates proposed scientific education and legal changes as means to hamper the clergy's negative influence, the radicals undertook acts of iconoclasm—such as burning or destroying Catholic symbols and images, parodies and public derision, etc.—that were intended to function as effective propaganda against religion. Herrera probably envisioned the hybridization experiment as an act of iconoclasm that would contribute to the destruction of the Catholic Church and, ultimately, any type of cult or religious beliefs.

In addition to atheism and Darwinism, England and Herrera added other novel motives and concerns in interpreting the hybridization experiment. The Mexican biologist's pamphlet incorporates an article written by England in which he poses the theory of an alleged correspondence or affinity between different types of primates and different human races. This theory—which was missing in Ivanov's original project—established that, due to its supposed biological affinity, it was convenient to cross the Black race with gorillas, the yellow race with orangutans, the alpine or "brachycephalic" race with gibbons, and finally the White race with chimpanzees.[17] According to England, "we have the program of crossing these four cousins (gorilla, orangutan, gibbon, chimpanzee) with the diverse races of men, hoping that they will produce four types of hybrids, which, if they are fertile, crossing them with the others, can give us a complete series of specimens, from the perfect anthropoid to the perfect man" ("tenemos el programa de cruzar estos cuatro primos [gorila, orangután, gibón, chimpancé] con las diversas razas de hombres, esperando produzcan cuatro tipos de híbridos, que, si resultan fecundos, cruzándolos con los otros, puedan darnos una serie completa de ejemplares, del antropoide perfecto al hombre perfecto").[18] It is evident that this alleged correspondence between classes of primates and human races relied upon a type of "scientific racism" that stipulated a racial hierarchy in biological terms, in such a way that gorillas' brutality and ugliness was linked to the presumed biological characteristics of the Black race, while chimpanzees' intelligence and sophisticated skills supposedly corresponded to the White race.[19] England predicted that, as a product of the crossing between the different types of human races and apes, the evolutionary and gradual chain that led to homo sapiens—and to the White race as the alleged culmination of the species—could be recreated entirely.

In his pamphlet, Herrera undoubtedly builds on the same biopolitical hierarchy with the aim of pondering the feasibility and consequences of the hybridization experiment. Forecasting the final characteristics of the hybrid, Herrera states that "its intelligence will be superior to that of the monkey and probably not much inferior to that of the hairy aborigines of Australia and the (New) Hebrides" ("su inteligencia será superior a la del mono y probablemente no muy inferior a la que tienen los velludos aborígenes de Australia y las [Nuevas] Hébridas").[20] In this quote it is clear that the Mexican biologist considered the non-White and non-Western races as a primitive human subtype akin to primates due to their inferior intelligence. In order to illustrate the plausibility of his experiment, Herrera even combined photographs of Indigenous men with chimpanzees to produce a projected image of what the hybrid might look like.[21] The "primitive" features of the Indigenous man at the right were emphasized to fit in perfectly with the chimpanzee pictured at the left when both photographs superposed in the middle. The presumption that the hybrid will look like and have similar traits to the so-called inferior races is consistent with the scheme "Evolution from the ether to the future superman" mentioned in Chapter 1, which places the "Malays" (a term referring to the Indigenous peoples of Southeast Asia) and "Blacks" as representatives of the lowest level of the "human" category and therefore closer to primates. This biological hierarchy is also evident in the article "Filosofía comparada. El animal y el salvaje" (1895), where Herrera argues that animals and "savage" peoples (so-called Redskins, Eskimos, Blacks, Gauchos, Apaches) possess in similar degrees basic faculties such as induction, experimentation, and ingenuity, and even higher attributes such as moral conscience, religion, charity, and love.

Like England, Herrera seems to suggest that evolution is a teleological trajectory that can be traced back to a specific point in the past for the purpose of recreating extinct life forms. While the viability of regressing a historical and contingent process that lasted hundreds of thousands of years was taken for granted, the debate arose in considering how many crosses between hybrids would be necessary to finally reproduce the characteristics of the extinct ancestor of human beings. Ivanov, and probably Herrera himself, thought that it would only take a few generations to obtain the desired result.[22] No consideration seems to have been devoted to the fact that the hybrid of the chimpanzee and the human, even if it perfectly displayed all the characteristics of the evolutionary ancestor of the human, would not live in the same biological or climatological circumstances of its original existence. Thus, it would not be possible to imitate the contingent conditions of adaptation that resulted, over thousands of years, in the homo sapiens species. Strictly speaking, the hybrid would represent a new living being or even a novel biological species that would conflict with and adapt to new and unpredictable living conditions.

Rather than supposedly being a resurrection of the extinct past, what was at stake in the hybridization experiment was a merging of temporalities that coincided in the present and pointed definitely toward the future. The "past" that remains in the present (humans of so called "inferior" races) would be biologically mixed with the absolute past of the human being (chimpanzees, gorillas, etc.) in order to produce a type of "future past": something that will happen in the future but is imagined as a reinstatement of the extinct past. In this sense, the hybridization experiment builds on the same assumptions as the project of synthesizing "supermen" analyzed in the first chapter: both initiatives are based on the idea that biological trajectories can be suspended at a certain point and voluntarily resumed to direct evolution toward a desired goal. This would only be conceivable on the basis of humanity's presumptive control of nature, understood as an external sphere upon which human beings can ultimately impose their will and shape accordingly.

Paradoxically, rather than reinforcing a mechanistic and teleological conception of evolution, Herrera conceived of the hybridization experiment as a way to counteract the "anthropocentric error" that arises when one presupposes "that man is the center of creation, or even, of what exists" ("que el hombre es el centro de lo creado, o mejor, de lo que existe") and ends up opening "an abyss between our species and the others, of plants and animals, and minerals" ("un abismo entre nuestra especie y las otras, de plantas y animales, y minerales").[23] On the contrary, obtaining a hybrid of ape and human "would demonstrate the unity of man and animals, as well as our bestial nature" ("demostraría la unidad del hombre y los animales, así como nuestra naturaleza bestial").[24] In other words, this experiment would allegedly provide factual bases for the Darwinian notion of evolution, which tends to destabilize the supposedly exceptional and unique character of the human species in nature. The controversy caused by the news of the experiment suggests that indeed the hybridization project was considered a biological intervention that transgressed hierarchies considered natural and immutable, particularly the presumed preeminence of the human being. This interpretation of the experiment would resonate with an animal studies trend that, building on Darwinian or neo-Darwinian positions, emphasizes the *identity* between animals and human beings by suggesting that many of the qualities considered particularly human (such as sentience or subjectivity) are shared by nonhuman animals.[25]

However, Herrera's aim of highlighting the identity between animals and humans is deeply problematic for several reasons. As already noted, although it explicitly intended to challenge the alleged exceptionality of human beings, the hybridization project constitutes in fact a utopia of biological control that assumes that humans can thoroughly know, manipulate, and reproduce the evolution of biological life. The plausibility of hybridization is based on the presumption that human beings, as the alleged summit of the evolutionary process, can preside over the biological sphere and even recreate naturally extinct life forms. Furthermore, Herrera's goal in proposing the identity between animals and humans is not to highlight the agency or creativity of chimpanzees, or their inherent rights as sentient beings, as proponents of this trend in animal studies usually argue (such as Tom Reagan or Peter Singer, among others). On the contrary, the hybridization project shows an evident disinterest or contempt for chimpanzees beyond their instrumental character as objects of experimentation, in such a way that hybridization (its assumptions, its method, and possible results) is justified from an anthropocentric point of view that ends up objectifying animals. Although he endeavors to challenge the hierarchy between human and animal that has predominated in Western modernity, Herrera not only reinforces it as the basic presupposition upon which his project rests, but also transfers that biopolitical hierarchy *within* the human species: non-Western races are considered *less than* human, fundamentally akin to animals due to their supposed biological proximity to primates.

At first glance, Herrera's hybridization project may seem like an eccentric initiative that responds exclusively to early twentieth-century interests and concerns. However, plans to resurrect extinct biological species have resumed in recent decades through three plausible methods: backbreeding, genomic transfer, and direct gene editing.[26] The first of these is surprisingly similar to the one that Ivanov and Herrera had devised: so-called "selective backbreeding" consists of artificially inseminating a current species that has the desired phenotypic characteristics and to do so until the progeny reproduce the physical attributes of the extinct species. This method has already been practiced by two groups of scientists with the purpose of "resuscitating" the auroch and the quagga.[27] The second method—genomic transfer—consists of inserting a cell nucleus of the extinct species, which had been preserved frozen, into the female ovum of a current species and then inserting it into her uterus to bring it to term. This method has led some scientists to claim that it is possible to "resurrect" species such as the Pyrenean ibex and gastric brooding frogs.[28] The last method—direct gene editing—is similar to the previous one: when it has not been possible to find a cell nucleus of the chosen species in good condition, scientists can edit the genome of an akin species to reproduce the desired characteristics and later insert the ovum into a uterus. Direct gene editing has been the selected method to attempt the "resurrection" of the wooly mammoth starting from the Asian elephant genome, and even the possibility of bringing back the Neanderthal, a hominid species that competed and mixed with Homo sapiens before it became extinct.[29]

Thus, Herrera's dream of reviving an extinct hominid seems to have remained intact by adapting to the age of genetic engineering. Contemporary so-called de-extinction projects, however, do not reflect the purpose of proving evolutionary theses but are rather underpinned by the objective of "re-wilding": recovering biodiversity lost due to anthropogenic reasons.[30] Some scientists are currently proposing that "de-extinction" is the ultimate way of respecting nature by compensating for the role played by humans in the extinction of many biological species. Paradoxically, they endeavor to restore a sense of faithfulness to natural processes through a radical intervention in the biological sphere. According to Sophia Roosth, "synthetic biologists' resurrection and de-extinction projects are fantasies of ultimate biological control: their interventions seek to produce wholly synthetic creatures that will stand in, counterintuitively, as semblances of 'untouched' nature."[31] Like Ivanov and Herrera in their historical moment, contemporary advocates of these plans do not seem interested in pondering the unintended consequences and ethical dilemmas that the implementation of their projects might entail.[32]

THE TOTAL RESURRECTION OF THE PAST

In addition to the hybridization project, Herrera envisioned other highly speculative plans to resurrect extinct life forms. Surprisingly, these utopian figurations resonate to some extent with current genetic resuscitation projects. Just as the recent discovery of the woolly mammoth's genetic information, frozen in the permafrost of Siberia for millions of years, opened up the possibility of replicating this extinct species, Herrera dreamed that future scientists would possess unimaginable technological devices that would be able to track down the particles that had belonged to past living beings in order to rebuild their old form. "Science," proclaims Herrera, "announces the total possible resurrection" ("La ciencia anuncia la resurrección posible total") in such a way that it becomes possible to imagine that future scientists will be able to resuscitate not only simple organisms—as plasmogeny was presumably already achieving—but also famous human beings like Victor Hugo: "if future wise men come to mold electrons, first to make atoms, molecules and then substances, proteins, protozoa, metazoans, superior beings, men or rather, descendants of the current ones, we will again get to see the famous lovers, and perhaps even the resurrected Victor Hugo, standing on the pedestal of lightning and singing sublime verses to the Great War" ("si los sabios futuros llegan a amoldar los electrones, primero para fabricar átomos, moléculas y enseguida substancias, proteínas, protozoarios, metazoarios, seres superiores, hombres o más bien, descendientes de los actuales, llegaremos a ver de nuevo los amantes célebres, y tal vez aun a Víctor Hugo resucitado, irguiéndose sobre el pedestal del rayo y cantando versos sublimes a la Gran Guerra").[33]

Herrera expanded on the idea that the same universal flows of matter have created and transformed everything that exists in the universe since the beginning of time. These flows of matter are neither created nor destroyed, they are only transformed: "it can be assured that the granitic rocks of the Archean Period, formed from 80 to 130 million years ago or more, can nourish with their potash the plants that currently feed us, in part (healthy vegetables), and, therefore, nourish our cells" ("se puede asegurar que las rocas graníticas del Período Arqueano, formadas desde 80 a 130 millones de años, o más, pueden nutrir con su potasa las plantas que nos alimentan actualmente, en parte [las sanas hortalizas], y, por lo mismo, nutrir nuestras células").[34] In reality, according to Herrera, nothing that exists or has existed disappears completely, as its atoms are transformed and become part of other material organizations.[35] From this it follows that all human beings that have existed remain in some way or shape in the present, since they have only "expanded and disseminated in nature, and their electronic elements are found in us and around us" ("se han dilatado y diseminado en la naturaleza, y sus elementos electrónicos se encuentran en nosotros y en nuestro alrededor").[36] Although, over the course of his research, Herrera assigned different names to these basic elements—ether, electron, molecule—the idea of a single and universal material element that constitutes the cosmos by disarranging and recomposing itself remained fundamental to his thought.

Herrera even went further in theorizing that this permanence of the past in the present was an essential trait of nature, both in the organic and inorganic spheres: "if heredity preserves the past in chromosomes, in species and individuals, universal life must also possess an inheritance that brings together the infinity of distances and times" ("si la herencia conserva el pasado en los cromosomas, en las especies y los individuos, la vida universal deberá poseer también una herencia que reúne el infinito de las distancias y los tiempos").[37] Just as the project of hybridizing apes and humans implied that evolution was a teleological and necessary trajectory, the idea of a material reconstruction of extinct beings builds on a mechanistic conception of nature that pervades in plasmogenic experiments. Both the notion that it is necessary to reduce systems to their most elementary components (atomism) and the assumption that all vital processes can be explained by means of physical and chemical laws, in addition to the need for an experimental methodology, are features shared by several mechanistic biologists of the first decades of the twentieth century.[38] In essence, Herrera and other biologists such as Jacques Loeb (1859–1924) assumed that if the components of a mechanism (such as an organism) are known and the laws governing the interaction of those components are also known, the outcome of the process can be predicted with confidence. However, as I will discuss in the next section, Herrera produced at the same time a theorization that destabilizes the mechanistic, teleological, and anthropocentric guidelines that underpin his utopian plans of resurrecting extinct life forms.

CHEMICAL ETHICS

Herrera formulated a radically materialist conception of nature that he named "ethereal philosophy" (filosofía etérea). This was a "chemical philosophy" (filosofía química) that sought to attune to the "ethics" of universal life. According to the Mexican biologist, the history of philosophy—from Plato's idealism to Comte's positivism to Bergson's or Caso's vitalism—is composed of "apparent or disguised metaphysicians" ("metafísicos aparentes o disimulados") who are incapable of explaining reality in ontological terms without resorting to immaterial or spiritual principles.[39] In opposition to the Western philosophical tradition that expands on unproved principles, at present "the only possible philosophers are chemistry professors" ("los únicos filósofos posibles son profesores de química").[40] The prospect of a comprehensive materialist philosophy must start from the existence of a "primordial chemical element" ("elemento químico primordial"), the ether, which unifies and ultimately accounts for everything that exists: "I reduce what exists to the ether, from the sun to consciousness, the self, pleasure and pain, will, memory, thought" ("Reduzco lo que existe al éter, desde el sol hasta la conciencia, el yo, el placer y el dolor, la voluntad, la memoria, el pensamiento").[41] In keeping with the scientific ideas of his time, Herrera conceived of the ether as a kind of "imponderable matter" ("materia imponderable") that inhabits the entire universe forming a medium or "infinite sea that encloses everything" ("mar infinito que todo lo encierra") and making possible the existence of all natural phenomena (physical, chemical, biological or astronomical).[42] The various organizations of the ether produce processes as dissimilar as the planetary systems and human thought: "We only see modalities of the ether, vibrations and condensations" ("Solo vemos modalidades del éter, vibraciones y condensaciones").[43]

Herrera developed the ultimate implications of ethereal philosophy in a pamphlet published in Argentina in 1919. If all natural phenomena share the same physical-chemical composition, then the same ethics must govern the behavior of atoms, inorganic matter, biological organisms, and human beings: "Morality is the law and order of atomic, mineral, animal or human colonies" ("La moral es la ley y el orden de las colonias atómicas, minerales, animales o humanas").[44] The ethics of nature is determined by the physical-chemical affinities and disassociations that prescribe the universal circulation of matter and energy: "morality explores its aspects (of matter and energy), legislates on its militant units, its physical-chemical affinity, love, union, brotherhood; its disassociations, hatred, revolution, social crime" ("la moral explora sus aspectos [de la materia y la energía], legisla sobre sus unidades militantes, su afinidad físico-química, amor, unión, fraternidad; sus disociaciones, odio, revolución, crimen social").[45] Herrera even proposes a basic definition that distinguishes "good" from "evil" in the context of chemical ethics: "Practice good, forgiveness, compassion, peace, to increase life; flee from evil, the diminishing of life, fulfill the geological and ecumenical mission of the entities" ("Practicad el bien, el perdón, la compasión, la paz, para aumentar la vida; huid del mal, disminución de vida, cumplid la misión geológica y ecuménica de los entes").[46] In other words, the good consists in "increasing" or favoring the universal movement and circulation of what exists, whereas the evil resides in "diminishing" or hampering the infinite process of associations and dissociations. Embracing this incessant transformation is the only possible ethics of nature and, therefore, of human beings: "Morality is immanent, eternal, universal, fatal, mineralogical, biological, unitary, mathematical, incorruptible, transcendental, integral, absolute" ("La moral es inmanente, eterna, universal, fatal, mineralógica, biológica, unitaria, matemática, incorruptible, trascendental, integral, absoluta").[47]

Building on this notion of chemical ethics, Herrera necessarily arrives at some general conclusions. For example, it becomes unavoidable to conceive of the inanimate or dead as an *internal* variant of life: "For science, death is life. A modality of what exists eternally, germ of new beings and stage of the circulation of matter, of the molecules animated by the same scorched breath of the great whirlwind, the Milky Way, where eternity walks around meditating" ("Ante la ciencia, la muerte es vida. Modalidad de lo que existe eternamente, germen de nuevos seres y etapa de la circulación de la materia, de las moléculas animadas por el mismo soplo abrasado del gran torbellino, la Vía Láctea, donde se pasea meditando la eternidad").[48] Thus, death is actually a manifestation of "good"—and not of "evil" as usually assumed—because in the end it allows for the continuation of the movement of matter and energy through space and time. In the same way, war—a traditional source of pain and sadness—is presented as a provisional and necessary stage in the endless series of unions and disintegrations. The "immanent justice of the Ether" ("justicia inmanente del Éter") presupposes that great human conflagrations—Herrera was surely thinking of the Mexican Revolution and World War I—serve the ultimate purpose of renewing the constant processes of nature: "From the macabre graves that populate the earth instead of men, other humanities will emerge. The buried or scattered bones will nourish the wheat, they will become the bread of the peoples. That is to say, dead peoples will become living peoples" ("De las fosas macabras que pueblan la tierra en vez de los hombres, surgirán otras humanidades. Los huesos sepultados o dispersos, nutrirán el trigo, se harán pan de los pueblos. Es decir, los pueblos muertos se convertirán en pueblos vivos").[49] Universal "war" in its chemical, physical, astronomical, or political manifestations constitutes, thus, a basic premise of chemical ethics.

Another fundamental premise of Herrera's philosophical conception is the idea that there was never a "creation" in the strict sense—one carried out by a transcendental force or Creator, as the various religions affirm—because matter and energy have always existed, and its varying organizations have shaped nature. Just as there was no first cause that originated universal life, there is also no expected result or endpoint of the infinite chain of natural transformations. Although it explains nature by means of laws or scientific systems, ethereal philosophy cannot decree a necessary point of arrival of universal life. Chemical ethics is only responsible for describing the trends or trajectories of matter and energy, but it does not prescribe their ultimate end. In contrast to the deterministic and teleological assumptions at the base of his projects of resurrection of the extinct past, Herrera argued that the constant transformation of nature is not governed by a transcendental plan or intelligent force that teleologically prescribes specific final results: "The only plan that fundamentally has followed nature has been this: to determine by increasingly complex means the incessant transformation of force.... And in effect, being is a transforming machine of energy" ("El único plan que fundamentalmente ha seguido la naturaleza ha sido éste: determinar por medios cada vez más complicados la transformación incesante de la fuerza.... Y en efecto, el ser es una máquina transformadora de la energía").[50] There is, then, a fundamental tension in Herrera's thought between a radically mechanistic and reductionist conception of nature (accompanied by a teleological Darwinism), and a notion of chemical ethics that elucidates the differentiation and incalculability of matter and energy becoming in the universe.

THE LIFE OF METALS

It is insightful to observe and analyze this unresolved tension in the Mexican biologist's speculations about what he calls "the life of metals."[51] Herrera elaborates on the studies of the Belgian engineer and metallurgical expert James Van Drunen (1855–1932) and the Indian doctor and biologist Sir

J. Chandra Bose (1858–1937), who carried out experiments to show the agentic traits of this material, one usually considered static and inert. By analyzing the microstructural characteristics of metals (metallography), these studies showed that metals transform their structure when impacted by pressure, temperature, and the passage of time. Bose applied stimuli to various kinds of beings (including metals and plants, as discussed in Chapter 3) and studied their molecular reactions, discovering in metals a remarkable sensitivity or ability to physically respond that was thought to be reserved for higher organic bodies. As Herrera summarized, the Indian biologist showed convincingly that metals have a "conductivity, similar to the nervous one" ("conductibilidad, semejante a la nerviosa"), that is, they are capable of producing electrical signals in response to external stimuli and conducting those signals in a similar fashion to the human nervous system.[52]

In addition to sensitivity and conductivity, metals reveal under careful scrutiny an agitated inorganic life that is akin to what Herrera observed in the behavior of colpoids. For example, if one tries to break a metal by means of a strong pressure, it can be observed that "the metallic molecules or granules have been transported and accumulated around the threatened point of stricture, and they come to harden and consolidate this weak part" ("las moléculas o gránulos metálicos se han transportado y acumulado alrededor del punto de estricción amenazado, y vienen a endurecer y consolidar esta parte débil"), which represents a surprising "call for reinforcement, this call for help in the face of danger" ("llamado de refuerzo, esta llamada de socorro ante el peligro").[53] When subjected to different temperatures, steel shows a diversity of molecular behaviors: "A cold-worked steel shows broken glass, without coherence; if it is heated to anneal, a repair is made, the crystals heal and reconstitute. By cooling, or by compression, the fixation of certain 'instinctive' and always identical movements is achieved" ("Un acero trabajado en frío, muestra cristales rotos, sin coherencia; si se le calienta para recocerlo, se hace una reparación, los cristales se cicatrizan y se reconstituyen. Por enfriamiento, o por compresión, se obtiene la fijación de ciertos movimientos 'instintivos' y siempre idénticos").[54] Furthermore, metals reveal other rudimentary phenomena such as disease (corrosion), memory (they retain acquired properties for a certain time even after the removal of the stimulus that produced them), death (molecular movement stops completely), fatigue (a gradually decreased reaction to a sustained application of stimulus), among others.

Metals display a wide range of behaviors, qualities, and affects: they can have a liquid quality that allows them to take various forms or they can present a seemingly unalterable firmness. They are also capable of going from one extreme state to the other: from fluidity to firmness or from being shapeless to adopting a specific form, for example. Even when they acquire an apparently stable state, they are susceptible to being transformed once more on the molecular or external levels. Metals are affected by factors such as heat and pressure through processes such as casting or molding, but they are also capable of affecting other bodies with their qualities such as color, weight, sharpness, or resistance. This is why Deleuze and Guattari suggest that the metallurgical profession first stumbled upon the notion of inorganic life through their close and constant interaction with metals: "what metal and metallurgy bring to light is a life proper to matter, a vital state of matter as such, a material vitalism that doubtless exists everywhere but is ordinarily hidden or covered, rendered unrecognizable, dissociated by the hylomorphic model."[55] While the hylomorphic model builds on the assumption that raw and inert matter is outwardly transformed by the imposition of an external and immaterial form, material vitalists such as Herrera assert that all matter displays a sense of agency marked by its inherent features of expression and action that have the ability to affect and be affected by other bodies.[56] Similarly, recent reflection on new materialisms has also focused on the life of metals to theorize the agency of matter.[57]

Herrera even contends that the world of metals already shows the principle of evolution that has determined the mutations of biological life. If, according to plasmogeny's hypothesis, organic life was shaped on the planet from inorganic elements, from this it follows that the inorganic world must have a natural tendency to evolve toward more organized and sophisticated states of matter. Herrera considers that the logic of nature is determined by universal and mechanical laws that determine the interaction of inorganic particles and their evolution toward biological life forms. These natural laws not only regulate the evolution of life on Earth, but also presuppose that biological life as we know it will be created on all planets where the appropriate physical-chemical conditions exist (that Herrera calls the "law of planetary lives"). This principle of the unity of the universe also prescribes that those attributes of biological life that are considered superior must already exist in a rudimentary form in the inorganic world. The vitality of metals—their behaviors of sensitivity, memory, defense, etc.—is convincing proof that there exists in the inorganic world the starting point that will lead to the creation of biological life and human beings through a very long series of physico-chemical transformations. Here, as in the project of human-ape hybridization, Herrera displays a mechanistic understanding of matter closely linked to a teleological framework of evolution.

However, the life of metals also reveals at the same time a contingent and differentiating movement that does not adhere to a strict teleology or predictability. When molten steel is rapidly cooled, one observes the sudden formation of what Van Drunen calls "temporary, numerous, and special combinations" ("combinaciones pasajeras, numerosas y especiales"), which act erratically and only behave predictably at certain temperature limits.[58] In fact, the extraction of constant laws is only made possible by the subtending and constant variability of matter: as Deleuze and Guattari put it "one may link the materiality's power of variation to laws adapting a fixed form and a constant matter to one another. But this cannot be done without a distortion that consists in uprooting variables from the state of continuous variation, in order to extract from them fixed points and constant relations."[59] In spite of their stable appearance, metals and the inorganic world showcase a "continuous variation" of matter in constant motion, which is the basis of matter's chemical ethics. The endless variation promotes in turn the appearance of instances of what the Epicurean philosopher Lucretius called "clinamen": a spontaneous and unpredictable deviation of an atom that, by initiating a new trajectory, establishes an unprecedented and incalculable chain of actions. The notion of the clinamen inaugurates in the Western philosophical tradition the prospect of an "aleatory materialism" that, like Deleuze and Guattari's or Herrera's, espouse the inherent variability and unpredictability of being understood as difference.[60] This philosophical tradition has been refashioned and expanded recently by new materialisms' emphasis on the lively, agentic tendencies of matter.

Based on these ideas, it can be argued that biological life constitutes a clinamen of inorganic matter: at some point in the history of the universe, continuous transformations of matter deviated and went on to form a new sphere of action and expression that we call biological life. Herrera suggests this precise argument by incorporating a quote from the composer Saint Saëns, who, upon carefully observing the Mont Blanc, exclaimed: "Organic life is an accident. What predominates is inorganic matter" ("La vida orgánica es un accidente. Lo que predomina, es la materia inorgánica").[61] Biological life might be considered a historical *accident* that undoubtedly represented a manifestation of "good," according to the criteria of chemical ethics, because it implied a remarkable moment of prolongation, differentiation, and creativity of matter. But the enormous variety of associations and disassociations of matter that exceed—and also undergird and pass through—biological life is in actual terms the most prevalent feature of the universe. If the inorganic is usually regarded in negative terms as something that "lacks" life or does not "achieve" vitality, Herrera's chemical ethics invite us to think that biological life might not be the final endpoint or the maximum ontological degree that all beings must aspire to reach.

In the same vein, chemical ethics invites us to ponder the infinite potentiality of inorganic life, which is capable not only of forming biological life, but also of *not* partaking in the organic world. As Elizabeth Povinelli has recently argued: "what if the miracle was not Life, the emergence of a thing with new forms and agencies of potentiality, but Nonlife, a form of existence that had the potential not merely to be denuded of life but to produce what it is not, namely Life? Nonlife has the power to self-organize or not, to become Life or not. In this case, a zero-degree form of intention is the source of all intention. The inert is the truth of life, not its horror."[62] The ability of inorganic matter to self-organize and create forms (morphogenesis) could be better understood as the *truth* of biological life, not its exclusionary opposite. In this way, chemical ethics reveals the complex and unpredictable behaviors of matter that underlie and destabilize biopolitical hierarchies and biologistic understandings of life.

THE VITALISM OF ANTONIO CASO

An early critique of Herrera's mechanistic and materialist tendencies appeared in philosopher Antonio Caso's *La existencia como economía, como desinterés y como caridad*. As mentioned in the introduction, Caso belongs, along with Alfonso Reyes, José Vasconcelos, and Pedro Henríquez Ureña, to the generation of Ateneo de la Juventud that undermined the prevailing positivism among the intellectual and political elites of the Porfiriato through the incorporation of notions akin to vitalism and spiritualism.[63] A fundamental objective of Caso and other members of his generation— especially Vasconcelos, as will be seen in Chapter 3—was to dismantle the predominant "biologism" that, in line with Mexican social Darwinists like Gabino Barreda, tended to account for all spheres of human activity by referring to an underlying biological substrate.[64] Caso and his peers, on the other hand, argued that the life of human beings could not be explained by purely biological theories, since human domains such as culture, religiosity, charity, or love transcend and even contradict biology. In *La existencia como economía, como desinterés y como caridad*—originally published as a short pamphlet in 1916, expanded in 1919 and again in 1943—Caso developed the basic features of his vitalist philosophical system marked by a Christian-inspired humanism.[65]

In the 1919 edition, while Herrera was developing his philosophical conception and his laboratory experiments, Caso includes a critique of plasmogeny and of materialist determinism in general to emphasize his own conception of life: "the experiments in plasmogeny or plasmogenesis that some contemporary researchers ingeniously carry out using the chemical components of life are nothing but imitations of form, never representations of function; and life does not consist in its form separated from its matter nor in its matter separated from its form, but in both merged together and composing organs, apparatus and organisms, more or less complex" ("los ensayos de plasmogenia o plasmogénesis que ingeniosamente realizan algunos investigadores contemporáneos valiéndose de los componentes químicos de la vida, no son sino remedos de la forma, nunca trasuntos de la función; y la vida no consiste en su forma separada de su materia ni en su materia separada de su forma, sino en ambas avenidas y realizadas en órganos, aparatos y organismos, más o menos complejos").[66] It is unsurprising that Caso singles out Herrera—without mentioning him—as representative of a materialist strand of biologism, since the founder of plasmogeny was one of the most renowned Darwinists in the country and had a long career in the scientific community. Caso, in contrast, was a young humanist philosopher who wanted to differentiate his own thought from the positivist biologism of the past generation. Whereas fellow member of the generation Alfonso Reyes speculates about plasmogeny's artistic implications—as seen in Chapter 1—Caso carries out a careful analysis of the assumptions underlying plasmogenic experiments in order to serve his own philosophical project.

Caso maintains that these experiments are constrained to recomposing matter to produce the form of life; they cannot create the vital function that is ultimately irreducible to neither matter nor form. The vital essence cannot reside in the form, because organisms transform their morphological aspects (they grow, reproduce, etc.) in response to environmental stimuli. Similarly, matter itself does not constitute the vital principle because living bodies are made of the same matter that exists in the inorganic world. The essence of life, then, cannot be reduced to the material or morphological components of organisms without missing out on what makes them truly alive: "Materialists do not see that which is characteristic of life because it is not material. In an imperfect sense, there is nothing in living substance other than matter and physical forces. It is all the same. But in the absolute and perfect sense, everything is diverse, because a new trivial cause has intervened: nutrition, the directing force as Claude Bernard used to say, the function" ("Los materialistas no ven lo característico de la vida por lo mismo que no es material. En un sentido imperfecto, nada hay en la sustancia que vive diverso de la materia y las fuerzas físicas. Todo es lo mismo.

Pero, en el sentido absoluto y perfecto, todo es diverso, porque ha intervenido una nueva causa trivialísima: la nutrición, la fuerza directora que decía Claude Bernard, la función").[67] Caso criticizes what he considers the blindness of materialist and reductionist biologists like Herrera who end up disregarding the essential differences between the inorganic and the organic, the inert and the living.

Thus, according to Caso, the essence that separates life from the inert is a specific *function* inherent to all living things, which is best defined as "perpetual nutritive movement" ("perpetuo movimiento nutritivo").[68] Nutrition represents "a new dynamic principle" ("un nuevo principio dinámico"), absent in inorganic matter, characterized by "the need for expansion, for conquest" ("la necesidad de expansión, de conquista").[69] If "life is an immanent purpose of hoarding" ("la vida es una finalidad inmanente de acaparamiento"), it is because the vital nutritive function consists of "hoarding" or "conquering" environmental resources with the aim of assimilating and integrating them into the always dynamic balance of organisms.[70] This "essential selfishness" ("egoísmo esencial") *is* biological life: "insatiable greed, selfishness, the spirit of domination. It makes the inert, the inanimate (and the animate as well) its own. It imposes its own stamp, shape, or formula on it. It tyrannizes it in order to exist" ("la avaricia insaciable, el egoísmo, el espíritu de dominación. Hace suyo lo inerte, lo inanimado [y lo animado también]. Le impone su sello, su forma o fórmula propia. Lo tiraniza para ser").[71] The vital function establishes a selfish exchange between the organism and the environment, guided by the economic principle par excellence: "a maximum of profit with a minimum of effort" ("el máximun de provecho con el mínimun de esfuerzo").[72]

The vital function plays a crucial role in Caso's philosophical system, because it represents an "orden autónomo" that even within the basic biological level, seems to *transcend* biology and point to higher levels of existence (art, charity, etc.).[73] In other words, whereas biologistic viewpoints pose biological systems as a sort of material foundation that determines and ultimately explains all realms of human existence, Caso strove to prove that in biological existence there is already a non-material, autonomous impulse that naturally generates a "vital excess" ("demasía vital"), a surplus of energy that is necessarily spent in non-biological, disinterested activities such as play or art.[74] Thus, in accordance with Vasconcelos's philosophical system discussed in Chapter 3, for Caso human beings represented the order in which life finally frees itself from its fundamental egoism to achieve the realm of (Christian) spirituality and charity, which is clearly an antibiological field that represents the antithesis of the economic principle by prescribing the maximum of effort with a minimum of profit.⁷⁵

Significantly, the vital function theorized by Caso is not found at the level of the elementary particle—as Herrera argues—but at the level of the functional system. In other words, the "perpetual nutritive movement" ("perpetuo movimiento nutritivo") can only be carried out by "organs, systems, and organisms" ("órganos, aparatos y organismos") and not by matter itself. Caso's perspective allows us to consider that organisms perform functions that exceed the simple sum or interaction of their elemental parts. This organicist perspective provides the basis for criticizing what Caso considers the mechanistic and simplistic conception of vital processes espoused by Herrera and the plasmogenists. This criticism becomes especially pertinent when considering Herrera's projects related to the creation of future or extinct life forms, which do not sufficiently account for the complexity and variety of contingent and unpredictable determinants that are involved in vital processes (from the atomic level, going through the levels of molecules, cells, tissues, organs, organisms, until reaching the level of populations, ecosystems, heredity, variation).

At the same time, as I argued earlier, Herrera theorizes a vital flow whose greatest destabilizing potential lies precisely in that it cannot be contained in organs, systems, organisms. Universal life is the flow of inorganic life that crosses biopolitical hierarchies and displays a "chemical ethics" based on the incalculable interactions established by matter. Caso's organicism ends up *separating* vitality from its material basis by conceiving an autonomous function that animates organisms: as he puts it, "there is a metaphysical principle—*hypotheses fingo*—an energy different from the physical-chemical, which acts in the world forming the whirlwind of life" ("hay un principio metafísico—*hypotheses fingo*—; una energía diversa de las físico-químicas, que actúa en el mundo formando el torbellino de la vida").⁷⁶ Unlike Caso, Herrera expands on the idea of material vitalism: a sense of vitality that does not consist in a metaphysical principle or a special function of organic matter, but in the active capacities—such as creativity, self-organization, and so on—that are inherent to the inorganic sphere. Furthermore, while Caso proposes that the purpose of existence is to transcend life's "essential selfishness" ("egoísmo esencial") and reach the strictly ethical spheres of selflessness and charity, Herrera goes so far as to formulate a radical materialism that does not postulate an ultimate purpose beyond the endless prolongation of the chemical ethics of matter. As I will discuss next, if Herrera's material vitalism was criticized in philosophical terms by Caso, it was also rejected—although for slightly different reasons—in the nascent field of biological sciences in Mexico.

BIOLOGICAL DISPUTES

In the 1910s and 1920s, Herrera's speculations conflicted with another conception of biology closely associated to its medical and utilitarian applications, which was propounded by some of Herrera's close collaborators such as Eliseo Ramírez Ulloa, Isaac Ocheterena, and Fernando Ocaranza. As Ismael Ledesma-Mateos has argued, the fundamental disagreements between Herrera and physicians such as Ocaranza and Ramírez ultimately determined the path that the institutionalization of biology took in Mexico and Herrera's own individual fate in this scientific field.[77] Ultimately, the viewpoints held by Ocaranza and Ramírez secured discursive and institutional hegemony and ended up displacing the materialist, evolutionist, and philosophical conceptions of Herrera. This manifested institutionally in the demise in 1929 of the Dirección de Estudios Biológicos spearheaded by Herrera, making way for the establishment of the Instituto de Biología at the Universidad Nacional. This Institute was directed by Isaac Ocheterena, a biologist who—much like Ocaranza—worked alongside Herrera in the Dirección and showed a critical stance toward his experiments and theorizations. Since 1929, Mexican biology embarked on a course away from plasmogeny and philosophical reflection and headed toward the less speculative field of medical applications.[78]

One can get a clear sense of the disputes and controversies within Mexican biology by reading Fernando Ocaranza's memoir *La tragedia de un rector* (1943). Ocaranza was a surgeon who served as head of the Departamento de Fisiología Comparada at the Instituto de Biología General y Médica, part of the Dirección de Estudios Biológicos. In his reflections, Ocaranza describes the intellectual atmosphere at the Instituto and the varying relationships that the collaborators had with Herrera's research: "Some of those who worked at the Institute felt drawn by the ideas of Don Alfonso, whether by irreducible suggestion or because of lack of culture; others, simulated conviction and enthusiasm, to suit his interests; but some of us, very few, kept ourselves within a certain reserve, in which case, we were looked upon with distrust and regarded as people who did not deserve to belong to an institution that would soon bring incomparable glory to our country" ("Algunos de los que trabajaban en el Instituto, sentíanse arrastrados por las ideas de don Alfonso, ya fuere por irreductible sugestión o por escasa cultura; otros, aparentaban convicción y entusiasmo, por convenir así a sus intereses; pero algunos, muy pocos, nos manteníamos dentro de cierta reserva, en cuyo caso, éramos mirados con desconfianza y tenidos por personas que no merecían pertenecer a una institución que pronto daría gloria incomparable a nuestro país").[79] Ocaranza makes it clear that he belonged to the latter group of people, because he did not agree with the final objectives nor the specific findings of Herrera's investigation.

Ocaranza portrays Herrera's theories and experiments as "fantasies" that "compromised the honor and seriousness" ("comprometía(n) el honor y la seriedad") of the postrevolutionary institutions.[80] According to Ocaranza, the artificial cells created by plasmogeny "are representations, but not realities, even though each element has a central condensation that looks like a nucleus and a peripheral part that represents a protoplasm" ("son representaciones, mas no realidades, a pesar de cada elemento, tiene una condensación central que parece un núcleo y una parte periférica que representa un protoplasma").[81] Only "by forcing the imagination a little" ("forzando un poco la imaginación"), Herrera and his disciples were able to identify signs of cells' vital functions.[82] In this way, Ocaranza's position is akin to Caso's when he affirms that "it is not the same to fabricate imperfect representations of a cell than the cell itself, in which the function derived and perfected by nutrition and inheritance stands out above all" ("no es lo mismo fabricar representaciones imperfectas de una célula que la célula misma, en la cual se destaca sobre todo, la función derivada y perfeccionada por la nutrición y la herencia").[83] That is, both Ocaranza and Caso consider that plasmogenic experiments cannot duplicate the essential functions of life and are limited to imitating the superficial aspects of cells such as their morphology and some chemical reactions—osmosis, oxidation, etc.—that are omnipresent in nature.

Ocaranza accepts that artificial cells are "forms of chemical equilibrium, as are equally those of living beings" ("formas de equilibrio químico, como lo son, igualmente las de los seres vivientes"), but adds that they constitute bodies that merely carry out short-lived chemical reactions: "instantaneous evolutions, or revolutions, one can assert; or very fast reactions at least; but in any case, the newly formed bodies will soon be left in a state of chemical indifference" ("evoluciones instantáneas, o revoluciones, puede afirmarse; o reacciones muy rápidas cuando menos; pero en todo caso, pronto quedarán los cuerpos recién formados en el estado de la indiferencia química").[84] Thus, Ocaranza recognizes—and, immediately afterward, conceals or dismisses—what Herrera would see as the inherent vitality of matter that comprise the chemical ethics of nature. Although artificial cells perform ephemeral chemical reactions that also exist in life, Ocaranza argues that living beings accomplish considerably more complex activities such as assimilation to nourish themselves and contain more stable forms of organization such as organs. This higher level of complexity and organization qualitatively separates living beings from cells manufactured by plasmogeny. Thus, even when he shares with Herrera the acceptance of the so-called "physico-chemical doctrine of life" ("doctrina físico-química de la vida") and even the understanding of life as an evolution of inorganic matter, Ocaranza refuses to accept that plasmogenic experiments can duplicate the fundamental characteristics of living beings, because reproducing in the laboratory the temporary conditions that created life on Earth is ultimately unattainable: "the evolution of inorganic matter to become living has had to last centuries and even if it only were this circumstance, it is lacking in laboratories, even to produce imitations of matter suitable for life" ("la evolución de la materia inorgánica para convertirse en viviente, ha debido durar siglos y aunque fuera esta circunstancia, falta en los laboratorios, así sea para producir remedos de materia con aptitudes para la vida").[85]

Ocaranza criticizes the definition of universal life that sustains Herrera's thesis that "life is the movement in the Universe" ("la vida es el movimiento en el Universo"), which was engraved on a wall of the museum Herrera directed. Ocaranza deconstructs the presuppositions of this definition in order to show that it is not an objective or strictly scientific concept: "it seems useless to me to say 'in the Universe', since such an expression forces us to suppose that something exists outside or beyond the Universe, and so the author of such definition who, surely, does not pretend to be a mystic, falls into mysticism by stating such a thing. In truth, such a frame of mind should not cause surprise: the materialists are at times as mystical as the spiritualists, no matter how much they deny it or do not want to understand it" ("me parece inútil decir 'en el Universo', ya que tal expresión obliga a suponer que algo existe fuera o más allá del Universo y el autor de tal definición que, seguramente, no pretende ser místico, cae en el misticismo al asegurar tal cosa. En verdad, tal estado de ánimo no causa extrañeza: los materialistas, resultan a las veces tan místicos como los espiritualistas, por más que lo nieguen o no lo quieran entender").[86] In short, just as Herrera himself alleged that the philosophical tradition was full of "metafísicos disimulados," Ocaranza accuses Herrera of hidden metaphysical or mystical thinking.

Ocaranza thus reprimands Herrera for the fact that his speculations go beyond the strictly scientific sphere to delve into the examination of the ontological constitution of the universe. As Ocaranza suggests, even if one removes the phrase "in the Universe" from Herrera's definition of "life is the movement in the Universe," the concept of life would end up completely equated with the physical notion of movement, rendering the term "life" meaningless "if there is another word that everyone understands better: movement" ("si existe otra que todo mundo comprende mejor: movimiento").[87] Then, by trying to encompass all cosmic phenomena, Herrera's philosophical definition of universal life would not only acquire mystical or metaphysical undertones, it would also end up losing all explanatory meaning in strictly scientific terms.

In his memoirs, Ocaranza extensively cites the article "La simulación en la investigación biológica" (1922) by Eliseo Ramírez Ulloa, a doctor and professor who published pioneering research in the 1920s on the classification of vaginal cells and laid the groundwork for what would become known as the Pap smear. Ramírez's article contains one of the staunchest criticisms of plasmogeny, referring to Herrera's experiments as a "pseudo investigation" ("pseudo investigación") that represents "an evil done to incipient Mexican science" ("un mal causado a la incipiente ciencia mexicana").[88] Ramírez laments that an illegitimate and baseless field of study such as plasmogeny found material support and legitimacy from revolutionary administrations. In his words, in Mexico there is a lack of infrastructure and of "honest, enlightened and capable researchers" ("investigadores honrados, ilustrados y capaces"), which generates a "confusion between true scientific work and pseudo work, as evidenced by the fact that official elements protect a series of weird discussions (plasmogeny), which are nothing other than simulation in biological research" ("confusión entre el verdadero trabajo científico y el pseudo trabajo, como lo demuestra el hecho de ser protegida por elementos oficiales una serie de discusiones raras [plasmogenia], que no son otra cosa que la simulación en la investigación biológica").[89] Like Ocaranza and Caso, Ramírez argues that the intention to artificially create life is a misguided purpose that is doomed to failure, because the vital function is such a sophisticated mechanism that by definition it cannot be imitated in laboratories.[90]

According to Ramírez, despite the fact that it is not guided by genuinely scientific purposes, the science developed by Herrera has managed to convince a "meager environment" ("medio raquítico") with an "insufficient culture" ("cultura insuficiente") to recognize and value authentic research: plasmogeny "reveals such a lack of knowledge of natural phenomena that it has made all people wonder, in a major psychic disorder, because all people assume that in the so-called scientific research there is, above all, good faith and not interests of another order" ("revela un desconocimiento tal de los fenómenos naturales que ha hecho pensar a toda la gente, en una alteración psíquica de importancia, porque toda la gente supone que en la investigación que se dice científica existe, ante todo, buena fe y no intereses de otro orden").[91] Both Ramírez and Ocaranza agreed that Herrera's delusions were based on a lack of expertise in the various scientific disciplines that plasmogeny sought to transform. Ocaranza attests that Herrera "confessed himself to lacking deep knowledge in physics and chemistry; besides from the fact that, according to him, they were not indispensable for the new directions that he was trying to impress on biology and, for this reason, those that he possessed in those sciences seemed sufficient to him" ("confesaba él mismo, no tener conocimientos profundos en física y química; aparte de que, según aseguraba, no eran indispensables para los nuevos rumbos que trataba de imprimir a la biología y, por tal razón, los que poseía de aquellas ciencias, le parecían suficientes").[92] While Ramírez affirmed that plasmogeny "does not understand anything about physics or microbiology" ("no entiende nada de física ni de microbiología"), Ocaranza believed that Herrera's attempts to imitate the human brain "revealed its complete and astonishing ignorance of the embryology of the cerebro-spinal axis" ("revelaban su completa y pasmosa ignorancia de la embriología del eje cerebro-espinal").[93] Thus, according to these doctors, the speculative and transformative nature of Herrera's thought was projected onto fields of knowledge—such as physics, chemistry, microbiology, or embryology— that the founder of plasmogeny neither mastered nor could understand.

At the same time, Ocaranza and Ramírez attribute Herrera's scientific incompetence to a personality trait that begged various interpretations: "either he was a simulator ... ; or he was a fanatic ... of his own ideas; or he suffered from a very particular state of mind that could be called eagerness for discoveries. Benevolently, I would accept the last explanation" ("o era un simulador ... o era un fanático ... de sus propias ideas; o sufría un estado mental muy particular que podría llamarse, afán por descubrimientos. Benévolamente, aceptaría yo la última explicación").[94] Thus, both doctors suggest that Herrera's work displayed traits of simulation or fanaticism, but the main factor driving his ill-advised investigations was what Ocaranza calls "the delusion of generalization and the eagerness to discover" ("el delirio de generalización y el afán de descubrir"), that is, the impulse to draw theoretical speculations of major significance from limited laboratory experiments and the ambition to explore and inquire into virtually unexplored realms of knowledge.[95] While physicians such as Ocaranza and Ramírez took for granted that biological research should have limited and pragmatic objectives, Herrera set out to tackle the most controversial issues and answer the philosophically important questions. Without a doubt, projects such as the synthetic creation of life or the resurrection of extinct life forms exceeded the interests and limited ambitions of the medical community. In this sense, Ocaranza and Ramírez highlight the fact that the utopian impulse that guided Herrera's thought was not consistent with, and even actively disrupted, the more constrained, practical aspirations of the medical sciences.

It is quite meaningful that Ocaranza and Ramírez use terms such as "fantasy" or "imagination" and phrases such as "representations, but not realities" to disqualify Herrera's experiments and philosophical speculations. Indeed, as I have argued in the first chapter, Herrera conceptualized his experiments as open-ended heuristic tools that could reveal unsuspected possibilities both for science and for perception and thought in general. Just as art can be understood as a representational or fictional tool for triggering thought, plasmogenic experiments created a kind of secondary reality that suggested new questions and paths for reflection. What Ocaranza called "the delusion of generalization and the eagerness to discover" was nothing more than Herrera's aspiration to follow unrestrictedly those unprecedented orientations that his experimental work opened up for all fields of knowledge and for the understanding of the universe. Whereas this led Herrera to pioneer in innovative fields of study such as the origin of life, it also led him to promote ideas or hypotheses that certainly, as his detractors claimed, were not always firmly based on solid knowledge. In the same way, this desire for generalization pushed Herrera to conceive philosophical ideas that went beyond the strict conceptualization of the scientific method as the basis of biological research and ventured into a radically materialist philosophy of nature.

Thus, Herrera's thought was equally attacked by physicians such as Ocaranza and Ramírez who questioned the legitimacy and usefulness of his theories, by philosophers such as Caso who disputed his radical material vitalism, and even by clericals like Emeterio Valverde Téllez, who rejected plasmogeny's materialist and atheistic tendencies.[96] In 1929, the criticism levelled against Herrera by his close collaborators led to the demise of the Dirección de Estudios Biológicos and with that he lost a great deal of institutional power and influence: from 1929 on "he [Herrera] was increasingly isolated from the academic establishment and became a minor player in Mexican national science."[97] In spite of that, from 1929 until his death in 1942, Herrera continued carrying out plasmogenic experiments in his private laboratory and circulating the results of his biological and philosophical research.[98]

THE BIOCOSMISM OF DIEGO RIVERA

Diego Rivera introduced in his own work some of the biocosmist themes and reflections developed by Herrera, including a general materialist outlook, an interest in the origin of life on Earth, the representation of universal life, the depiction of humanity as a cosmic agent, and the utopian potential of technology in the task of creating an improved biological humanity inhabiting the universe. In 1934, Rivera completed his mural painting *El hombre controlador del universo*, also known as *El hombre en el cruce de caminos*, at the Palacio de Bellas Artes in Mexico City. More than ten years after *La creación*—discussed in the introduction—the now famous muralist reconsidered some of the biocosmic themes tackled in his initial mural.

El hombre controlador del universo can be divided into three sections: the left section depicts salient features of capitalist social life, including a group of soldiers with gas masks and rifles marching into battle, policemen in horses suppressing a protest by unemployed workers, and idle elites playing cards and drinking cocktails. On the opposite side, the mural shows distinctive images of a communist society such as an orderly parade of male and female workers with red flags, a group of healthy female athletes moving in harmony, and the Bolshevik leader Vladimir Lenin joining hands with a multiracial group of soldiers and workers. In the middle section, located "at the crossroads" between the opposing social alternatives, appears a figure of a White worker operating a monumental, intricate arrangement of technological mechanisms for agricultural, scientific, and industrial purposes. A telescope and a microscope open up views of the macroscopic and microscopic worlds, which are depicted in four elliptical shapes radiating from the center of the mural. In addition, the worker possesses the ability to produce atomic energy, portrayed by a large hand controlling a sphere with atomic particles, one of them undergoing a process of nuclear fission. The worker is also manipulating machinery that enables him to exploit the surrounding natural resources by irrigating the crops and mining the geological strata pictured at the bottom.

Scholars have traditionally analyzed *El hombre controlador del universo* by focusing on the crucial political issues Rivera was engaging in through the use of visual and compositional elements alluding to class struggle, US capitalism, Soviet communism, and related issues.[99] Recently, Catha Paquette has explored how Rivera's mural addressed the most pressing

FIGURE 2.1. Diego Rivera's *El hombre controlador del universo*. Provided by Schalkwijk / Art Resource, NY. Diego Rivera, © 2024 Banco de México Diego Rivera Frida Kahlo Museums Trust, Mexico, D.F. / Artists Rights Society (ARS), New York

debates of the early 1930s, a historical period marked by the Great Depression of 1929, the proliferation of workers' protests and movements, the communist debates between the followers of Trotsky and Stalin, and the consolidation of the postrevolutionary regime in Mexico.¹⁰⁰ This type of reading is based on the dialectical relationship between the left and right sides of the mural, which depict the larger ideological struggle between capitalism and communism, and how the central figure of the worker has the political and technological power to choose the future society. Accordingly, Paquette accounts for the central macrocosmic and microcosmic motives—portrayed in the elliptical shapes—in the context of a reading that highlights Rivera's political purposes of asserting communist ideals such as egalitarianism, social stability, physical health, and technological expertise.¹⁰¹ Similarly, Mary Coffey emphasizes how "modern science and technology were folded into the historical opposition between fascism and communism."¹⁰² Drawing from these readings, I aim to delve in more detail into the middle section in order to restore the strictly *biocosmic* implications of this work, alluded to by the mural title itself.

In the same vein as Herrera's poetry, the middle section depicts the full range of expression of universal life at the cosmological, biological, geological, and atomic levels: from celestial bodies to plants, human beings, unicellular organisms, rock formations, and atoms. Humanity is situated as both a product of cosmic energy and a privileged arbitrator of its development: in line with Herrera's technological imagination, the human being in this mural is the ultimate *controller* of universal life through the use of technology.¹⁰³ The figure of the worker thus condenses Rivera's distinctive "technological utopianism," since he has the power and the transcendental responsibility of making a choice between life-sustaining or death-driven processes, both derived from the cosmic energy and pictured dialectically in the elliptical shapes.[104] On the left, coinciding with capitalist society, the ellipses contain a macroscopic image of a sun illuminating a lifeless white planet or star, as well as a microscopic view of disease-causing microbes (Trypanosoma, gonococcus, meningococcus, and spirochetes, among others) attacking cells, tissues, or organs, including a dissected penis and testicles contaminated by bacteria at the upper end of the ellipse.[105] On the right side, paired with the thriving communist society, the elliptical shapes contain a macroscopic glimpse of a giant red star—a symbol associated with communism—spiral nebulas or galaxies, and other celestial bodies such as planets, moons and asteroids; and a microscopic view of healthy cells, vessels, and organs, including a cross section of a breast showing milk ducts, a testicle in the process of developing sperm, and a massive ovary releasing an egg close to the image of an unborn child.[106] At the intersection of these ellipses showing cosmological and microbiological forms of vitality is a nursing infant, which seems to be sucking cosmic energy directly from the source, and represents the creation of a new biological humanity. Next to the infant, making a clear political association, the image of Lenin holding hands with a group of workers extends and consolidates the life-sustaining processes in human society.[107] In other words, what is at stake for the central worker manipulating an array of technologies is the future not only of human societies but more importantly of the cosmos as a whole: the choice between capitalism and communism is ultimately a choice of *cosmic* scale between the proliferation of universal life or its decline and expiration.

It can be argued that in *El hombre controlador del universo* Rivera came back to the biocosmic framework espoused by *La creación* in order to adjust it to the pressing issues of the time and his current political convictions. The two murals share the same structural basis: whereas the middle section gives an account of the diversification of the cosmic vitality with the human being occupying the central place, the left and right sides feature the ways in which human society might remain attuned to the path of universal life. Yet there are also important differences that speak of the transformation of the political climate and of Rivera's viewpoints. Whereas in *La creación* the cosmic energy is depicted in an abstract manner as a blue semicircle in the sky, in *El hombre controlador del universo* the same energy is symbolized in concrete terms by the massive potential stored in subatomic particles and the possibility of harnessing it, which was a promising prospect in the early 1930s. Accordingly, the central figure of the worker in the 1934 mural is portrayed as having the technological means to maximize the energy and resources stored in nature, in contrast with the nude figure with open arms—lacking any technology and social class—occupying the center of the 1921 mural. On both sides of the central human being, *La creación* depicts allegorical figures representing the importance of the arts and other spiritual faculties—but not of technology—in the process of ascending toward the primary cosmic energy. In contrast, *El hombre controlador del universo* depicts on the two sides of the central worker two antagonistic modes of production and social life, which condense opposing tendencies of universal life and its probable future outcome. In this way, while both murals provide an image of the progression of the cosmic vitality, at the same time they present vastly different conceptions of the specific ways in which humanity might play a crucial role in its future development.

The differences between Rivera's brand of biocosmism in the two murals can be accounted for by his changing artistic and intellectual motivations at the two points in time. In 1921, having just returned to his home country after a long stay in Europe (1907–1921) studying art history, Rivera had experienced little of Mexico's social and political climate—he visited the country briefly in 1910, witnessing a Zapatista battle—and his political beliefs were vaguely revolutionary.[108] Commissioned by José Vasconcelos to paint the walls of the National Preparatory School auditorium, Rivera was largely influenced by the early biocosmic ideas of his patron Vasconcelos, which will be discussed in more detail in Chapter 3. By the early 1930s, Rivera had strengthened his commitment both to artistic innovation and to Marxist political philosophy, which would result in the anti-Stalinist, anti-traditionalist stance espoused in the 1938 "Manifesto for an Independent Revolutionary Art." An important step in Rivera's process of artistic and political self-definition was his trip to the Soviet Union in 1927, during which he became involved with the avant-garde group October, which was at odds with the cultural policies backed by the Stalinist regime.[109] During his brief stay, Rivera was able to witness not only the beginnings of the restrictive Stalinist environment—which would become the probable cause of his sudden departure from the country in 1928—but also the impressive organization of the proletariat exemplified by the parade to mark the tenth anniversary of the Russian Revolution. Rivera did not miss the opportunity to make sketches of this magnificent parade, which later made their way into the right side of *El hombre controlador del universo*.[110]

It is probable that during his stay in Moscow Rivera came into contact with the well-established tradition of Russian cosmism, which was inaugurated by the late nineteenth-century thinker Nikolai Fedorov and thrived after the Revolution of 1917. Fedorov's influential philosophy promulgated a basic utopian figuration—namely, the prospect of establishing the technological and sociopolitical conditions that would allow humanity both to resurrect all past generations of human beings and to colonize the whole cosmos—which resonated intensely within the groups of avant-garde, revolutionary artists, and intellectuals of the 1920s and 1930s. Some of the intellectuals inspired by Fedorov, such as the Immortalist–Biocosmist collective led by Alexander Sviatogor, believed that the Russian Revolution and other class struggles around the world had set the stage for the achievement of their fundamental desires: immortalism and interplanetarism.[111] Not unlike Rivera's stance in his mural at the Palacio de Bellas Artes, for Sviatogor the consolidation of a communist society was a project of *cosmic* repercussions that would radically transform the fate of humanity and the universe as a whole. Even if fully-fledged cosmist thought was developed only by a handful of intellectuals and scientists—including Sviatogor, Vernadsky, Muraviev, Tsiolkovsky, Chizhevsky, among others—cosmism-inspired themes and perspectives were widespread in the artistic circles of the time, which were eager for radical, experimental philosophies. Conceivably, by becoming involved with avant-garde artists such as Mayakovsky, Rivera could have assimilated cosmist conceptions that well suited and supplemented his earlier cosmological ideas presented in *La creación*.

In addition to artistic circles, during his stay in the Soviet Union Rivera also established relationships with the scientific community by delivering a lecture titled "La revolución y la ciencia" at the Congress of Scientists.[112] He thus became acquainted with cutting-edge biological theories, particularly Alexander Oparin's theory of the origin of life on Earth, which—entirely like Herrera's plasmogeny—was derived from a materialist, evolutionist framework.[113] In the 1920s, Oparin and British scientist John Haldane independently arrived at the same working hypothesis: that the first organic molecules indispensable for living organisms must have been naturally synthesized as a result of the reaction between inorganic molecules and energy from lightning or the sun on the primitive Earth. Forming a sort of "prebiotic soup," these primordial organic molecules would have acted as the basic building blocks for the production of increasingly complex molecules and eventually organisms. In 1953, Harold Urey and Stanley Miller carried out an experiment that sought to verify the Oparin–Haldane theory: they simulated the conditions allegedly found on early Earth, including the presence of inorganic molecules such as methane and ammonia in the presence of repeated electric discharges, and after a few weeks they effectively managed to synthesize significant organic compounds.[114] The success of this

FIGURE 2.2. Diego Rivera's *El agua, origen de la vida*. Provided by Schalkwijk / Art Resource, NY. Diego Rivera, © 2024 Banco de México Diego Rivera Frida Kahlo Museums Trust, Mexico, D.F. / Artists Rights Society (ARS), New York

experiment attracted notable scientific and popular attention, consolidating the Oparin–Haldane theory as the basic framework for future research on the origin of life and practically dismissing concomitant theories such as Herrera's plasmogeny.[115]

Between 1950 and 1954, while the Miller–Urey experiment was gaining attention in newspapers around the world, Rivera painted the mural cycle *El agua, origen de la vida en la Tierra* at the Cárcamo de Dolores in Chapultepec forest, which served as the reception point of a monumental water supply system diverting the Lerma River to Mexico City. Originally conceived of as an underwater mural, this work explores the crucial importance of water in biological, socioeconomic, and political terms. The mural appears "like water, to spill out of a central tunnel" and fill the large reservoir that constitutes the main component of the work.[116] At the center of the reservoir, Rivera vividly illustrated Oparin's theory of the origin of life through a microscopic view of a group of primordial cells duplicating themselves and creating multiple individual cells.[117] A blue electrical discharge crosses the full microscopic field of vision, suggesting that these primordial cells were formed by the reaction of energy and inorganic compounds. The entire floor of the reservoir is filled with primitive organisms—polychaeta, heliozoa, cyanobacteria, and giardia, among many others—that gradually transform into fish, amphibians, reptiles, and eventually human beings depicted on the walls of the tank. As Rivera himself puts it, the mural portrays "a spark of electricity that animates the minerals in suspension and the first living cell is formed; the cell is subdivided and replicating in this way, it forms colonies, first microorganic and then more and more complicated, little by little it integrates into the plant and animal kingdom until it culminates in the human vertebrate" ("una chispa de electricidad que anima los minerales en suspensión y se forma la primera célula viva; ésta se subdivide y siempre reproduciéndose así, forma colonias, microorgánicas primero y después, de más en más complicadas, poco a poco se integra al reino vegetal y animal hasta culminar en el vertebrado humano").[118] Above the water level, which thus clearly demarcates the boundary between the socioeconomic and biological levels, Rivera painted the diverse physiological, economical, and recreational uses of water through images of workers and engineers providing drinking water to a thirsty population, a family watering an orchard, and children swimming.

El agua, origen de la vida en la Tierra delves into some of the biocosmic concerns explored by Rivera's work discussed earlier. In the vein of *El hombre controlador del universo*, the mural at the Cárcamo de Dolores focuses on the microscopic proliferation of vitality and, to a lesser degree, on the fundamental place of technology and engineering in the task of controlling the natural world. As in *La creación*, *El agua, origen de la vida en la Tierra* provides an image of the creation and diversification of life culminating in humanity, which is considered in its essential affinities and differences with mineral, vegetal, and animal beings and processes. Moreover, this mural grants similar importance to the role played by energy in the progress of cosmic vitality: whereas in his previous works this aspect is depicted as a blue semicircle in the sky, and as nuclear fission, it is now portrayed by the electrical discharge in the center of the reservoir. Perhaps what makes the mural at the Cárcamo de Dolores stand out in the context of his previous work is how Rivera decided to enhance the viewer experience by integrating and harnessing the conditions of the setting. Particularly, Rivera was hoping that the images in the mural would acquire a special brightness and movement—a sort of liveliness—by virtue of being under water: viewing the mural from a high vantage point, the spectator would experience the lengthy evolution of life on Earth being "brought to life" by the undulating movement of water, allowing them to trace humanity's history back to its inorganic origin. In the same vein as Herrera's aspirations for his own experiments, the viewer of this mural would therefore embrace their affiliation with universal life, while also accepting their privileged role in the task of commanding the universe.

It is not known whether Rivera engaged directly with Herrera's research on the origin of life, although it is hard to believe he was not at least aware of it. But beyond this point, it is clear that both Rivera and Herrera developed a strand of biocosmism that was firmly based in a shared set of concerns: the examination of universal life in materialist terms (from the inorganic to the organic spheres, and from the microscopic to the macroscopic worlds), the reflection on the role of humanity both as product and controller of universal life through technology, and lastly a shared utopian outlook that involves the destiny of the universe at large. While this set of biocosmic themes reinforces an anthropocentric framework closely linked to materialist and mechanistic conceptions, Herrera's theorization of chemical ethics seems to put pressure on this framework by emphasizing the unpredictable, complex behaviors of universal matter.

CONCLUSIONS

This chapter has explored the tension that runs through Herrera's utopian plans for resuscitating bygone forms of life. On the one hand, these plans relied upon a mechanistic and teleological conception of evolution that strived to highlight the identity between humans and animals in order to strengthen a secular and scientific understanding of nature. Rather than construing apes as subjects of ethical consideration, the project of hybridizing humans and apes by artificial insemination reinforced an anthropocentric control of nature and even introduced biopolitical hierarchies within humanity. Herrera attributed utopian potential to science and technology in the process of controlling and enhancing the whole universe, which is a common theme in Rivera's biocosmist works. On the other hand, Herrera developed a conceptualization of material vitalism as an attunement to the vital features of all kinds of matter and its contingent and unpredictable becoming. The chemical ethics of matter prescribes that "good" constitutes the continuing prolongation of physico-chemical reactions, while "evil" resides in the sudden arrest of these unpredictable transformations. Herrera's material vitalism represents a theorization that paradoxically undermines his own projects of total reconstruction of the biological past, opening up a contingent and open-ended understanding of nature. Whereas philosophers like Caso and biologists like Ocaranza rightfully contested Herrera's mechanistic views, they also largely neglected and dismissed material vitalism's wager of destabilizing traditional anthropocentric and organismic approaches.

In 1941, one year before his death, Herrera published a brief article titled "Una vida dedicada a crear la vida" in the Argentinean journal *Hombre de América*. Perhaps anticipating the end of his life, Herrera provided a summary of his lifelong research and insisted once again on its widespread philosophical, scientific, and medical implications for humanity. Sidelined by the Mexican academic establishment since 1929, Herrera asserted that even if "it is clear that (plasmogeny) concerns the greatest interests of humanity" ("es evidente que [la plasmogenia] atañe a los intereses más grandes de la humanidad"), people still tend to overlook and even actively obstruct his research: "It is deplorable that Plasmogeny is not equipped with immense laboratories and all the elements to propel it rapidly toward its final triumph" ("es deplorable que la Plasmogenia no sea dotada de inmensos laboratorios y de todos los elementos para impulsarla rápidamente hacia su final triunfo").[119] Herrera's research has concerned itself with approaching questions of major significance concerning life: "First and foremost, why do you live? what for? since when? What does life consist of, what is its origin, what is its physical-chemical basis, when did it appear on the planet when conditions were favorable?" ("En primer lugar, ¿por qué se vive? ¿para qué? ¿desde cuándo? ¿en qué consiste la vida, cuál es su origen, cuál su base físico-química, cuándo apareció en el planeta cuando las condiciones le fueron favorables?").[120] For Herrera, "a life devoted to creating life" implied grasping and engaging with the material basis of biological life, but also with the ceaseless transformations of inorganic life on Earth, which constitute the only end goal of nature and the true essence of life. Herrera died suddenly in 1942 while working in the laboratory he had built at home.[121] Next to his inert body, one could conceivably observe the colpoids and other artificial beings squirming lively under the microscope.

CHAPTER 3

José Vasconcelos

Botanical Ethics and the Cosmic Race

The intellectual career of José Vasconcelos has been thoroughly studied, from his institutional initiatives that left a decisive mark in postrevolutionary Mexico, to his highly regarded autobiographical works such as *Ulises criollo* (1935), to his spiritualist and utopian project of advancing the future fusion of all human races. While the political implications of all these facets of his career have been scrutinized—for example, the eugenic assumptions that underlie his miscegenation proposal—it is unquestionable that his life and work are obligatory references within Mexican and Latin American intellectual traditions.[1] In Vasconcelos's extensive corpus, his strictly philosophical texts have perhaps received the least critical attention and, in general, they often give rise to divergent assessments: José Gaos, for example, highlights the originality of his thought, while Samuel Ramos considers it a "work more of the imagination than of the intellect" ("obra más bien de la imaginación que del intelecto") that does not show the necessary thoroughness of truly philosophical endeavors.[2] In any case, Vasconcelos's philosophy is studied almost exclusively as "antecedent of filosofía de lo mexicano" ("antecedente de la filosofía de lo mexicano") or, more broadly, as a contribution to what has been called "the thinking of Nuestra América" ("el pensamiento de Nuestra América"), that is, the tradition of anti-imperialist Latin Americanism that strives to reaffirm a cultural and political opposition to metropolitan centers of power.[3] Vasconcelos's philosophical speculations that do not touch upon such identitarian issues often go unnoticed or are characterized as a mere eccentricity proper to the author's heterodox personality. His philosophical ideas about animals and plants, in particular, are rarely analyzed in detail and are mostly dismissed as insignificant or superfluous in overviews dedicated to his thought.[4]

This chapter explores Vasconcelos's understudied theorization on nonhuman ethics and his overall engagement with the notion of universal life. I will focus on Vasconcelos's particular conception of life across the cosmos and the ways in which nonhuman agents—minerals, animals and above all plants—display an ethical behavior. Particularly, I will delve into the model of ethics developed by Vasconcelos and what place horses and plants occupy in relation to human beings in his ethical scale. An examination of Frida Kahlo's depiction of human-plant hybrids and her cosmological ideas will lay the groundwork for studying Vasconcelos's theorization of plant consciousness and plant collectivity. As with the case of Herrera's thought, I will show that there is a structural tension in the philosophical works of Vasconcelos: a contradiction between the notion of (human) consciousness as a hierarchical product of human beings and the idea of plant consciousness as a close interdependence between organic and inorganic elements. Then, I will investigate which kind of "community" Vasconcelos admires in the life of plants, particularly in the jungle ecosystem. I will then examine how these philosophical ideas about plants and botanical ethics shed new light and cast an innovative image of one of Vasconcelos's best known texts, *La raza cósmica* (1925). Throughout the chapter I establish a dialogue between Vasconcelos's ideas and the interests of animal studies and plant studies, which contribute to destabilizing the anthropocentric notion of "the human" as a life form endowed with ontological, biological, and political privileges.

UNIVERSAL LIFE

La revulsión de la energía: Los ciclos de la fuerza, el cambio y la existencia (1924) is a short cosmological text that not only contains in a nutshell Vasconcelos's philosophical system—later elaborated in three volumes: *Tratado de metafísica* (1929), *Ética* (1932), and *Estética* (1936)—but also anticipates the main ideas of *La raza cósmica* (1925). In *La revulsión de la energía*, Vasconcelos gave an account of the creation and ongoing transformation of the whole cosmos by examining the transmutations of a primordial energy or force.[5] He was thus attempting to lay the foundation for a philosophical system that would contest accepted positivist frameworks in Mexico. As Abelardo Villegas has suggested, Vasconcelos endeavored to "oppose positivism point by point: in cosmology, theory of knowledge, axiology, and the philosophy of the history of the New World" ("oponerse al positivismo punto por punto: en cosmología, teoría del conocimiento, axiología y filosofía de la historia de América").[6] He pursued this project mostly by grounding his cosmological reflection, as well as most of his early philosophical texts, on Henri Bergson's vitalism in significant ways.[7] As Vasconcelos himself stated at a conference devoted to the French philosopher, Bergson's philosophy provided him with an ontological framework in which being was not conceptualized as an abstract or formal idea—as in Plato, Hegel or phenomenology—but as a concrete, dynamic energy, force, or potency that has multiple actual manifestations.[8] In contrast to the conceptualization of "essences that are fictions that do not represent living reality" ("esencias que son ficciones que no representan realidad viva"), Bergson afforded him the opportunity for a dynamic understanding of the wide-ranging realities and processes of the universe, from the celestial to the inorganic, the biological, and the human spirit.[9] As in the case of Caso's philosophy discussed in Chapter 2, Vasconcelos was keen to theorize a primary energy or force that, while being identifiable even in the most basic particle or organism, tended to *transcend* those elemental realms in search of higher states of being (the aesthetical or spiritual).[10]

Drawing on Bergson, Vasconcelos argues in *La revulsión de la energía* that the primordial energy has endured successive "revulsiones" or upheavals that, unlike the gradual changes of evolutionism, should be conceived of as abrupt transformations that not only rearrange existing phenomena, but above all *create* new potentialities of being. In this way, Vasconcelos implicitly subscribes to Bergson's theorization of creative evolution, which puts forward transformation not as a result of gradual adaptive changes, but as a process of genuinely innovative, unpredictable variation. For Vasconcelos there have been three massive cycles that cover the whole history of the universe: first, the cosmic cycle, encompassing the formation of nebula and all the planets; second, the planetary cycle that entails the appearance of celestial bodies including natural phenomena such as "the chemical life of minerals; the silent life of plants and the tumultuous life of species" ("la vida química del mineral; la vida silenciosa de las plantas y la vida tumultuosa de las especies"); and lastly, the proper vital cycle, which through its three periods—the material, the intellectual, and the aesthetic/spiritual—has engendered human beings capable of glimpsing the next and final cycle.[11] When the primordial energy reaches its final stage, Vasconcelos envisions that "the emanations of worlds will suspend their process. They will no longer be necessary because it is precisely the emanation that must depend on the effort of germs to expand their faculties, and multiply their experiences in order to flood into the infinity of the divine existence" ("las emanaciones de mundos suspenderán su proceso. No serán ya necesarias porque precisamente la emanación ha de depender del esfuerzo de los gérmenes por ensanchar sus facultades, y multiplicar sus experiencias con el fin de anegarse en la infinitud de la divina existencia").[12] In other words, clearly presupposing a "teleological conception of the universe" ("una concepción teleológica del universo"), in the final cycle the creative force will reach a full convergence with itself and will continue its unbounded movement in divine spheres.[13]

Vasconcelos proposes that the reality we live in—the universe that may be comprehended through perception and reasoning—constitutes only a small part of the wide array of beings and realms created by the primordial energy, which will also continue its creative movement in the future until it achieves a definitive elevation toward the Absolute/spirit. In other words, departing from the scientificism of the past generation, for Vasconcelos philosophy is not simply interested in understanding this world rationally, but more importantly is concerned with grasping the "inherent laws" ("leyes íntimas") and "essential processes" ("procesos esenciales") of the primordial energy through intuition.[14] By doing so, the philosopher may glimpse the imperceptible beings and worlds that currently coexist in the universe while also visualizing the dynamism of energy that outlines a sharply transcendent path. According to Vasconcelos, "Thus, a multitude of universes that we do not know coexist, because we lack organs that could insert us into the various environments" ("coexisten así multitud de universos que no conocemos, porque nos faltan órganos que pudieran insertarnos en los varios ambientes").[15] He speculates, perhaps informed by his theosophical readings, that there might exist other kinds of beings scattered throughout the universe in other planets and stars: beings whose physical composition might not even be perceived by human senses.[16] These beings are simply other experiments created by the primordial energy in its process of increasing perfection and elevation toward the Absolute/spirit. These unknown beings might be located in different phases of energy's path: while the moon could possibly be populated by inferior beings, the sun and other stars might be home to "beings of a spiritual constitution so solid that they inhabit the fire and take advantage of extreme light to penetrate the mysteries of the Universe" ("seres de una contextura spiritual tan firme que habitan en el fuego y aprovechan la luz extrema para penetrar los misterios del Universo").[17]

Even when some of these unknown beings might be considered inferior or superior to humanity, all of them—including human beings—are part of the same cosmological cycle in the sense that they have not reached the final phase of elevation of energy. Not unlike Herrera, Vasconcelos argues that between all existing phenomena—say, between minerals, plants, humans, and the inhabitants of the sun—there are only gradual differences, not essential distinctions. In the end, everything that exists is an emanation from a single creative force: "in our own conception and because it is a monistic dynamism, from one extreme to the other, the differences among nature and its classifications are a matter of degree and not radically primordial" ("en nuestra propia concepción y por ser ella un dinamismo monista, desde un extremo a otro, las diferencias de naturaleza y sus calificativos son cuestión de grado y no de raíz").[18] What Vasconcelos calls universal life is composed of all the entities and worlds—from cosmic space to earthly beings to the unknown beings populating the universe—that the creative force has engendered and will engender in the future:

> We see in time, series of events, processions of beings that come from a confusing thing that one does not know whether it is force or matter, but it appears as formless nebula or as already condensed matter, and it is distributed in fragments, and constitutes stars, and creates forms, not as something that disperses, but rather as something that becomes organized. And behind the form things and creatures appear. All brushed by a gentle touch that produces bright radiance, stars, music, and dance. Our eyes are dazzled, our ears are intoxicated, our hearts are delighted. The ear made for the winds and rumblings of the earth, picks up the rough vibrations; but the infinite rhythm permeates our souls and renews them with happiness and glory. Universal life leaves us perplexed.

> *Vemos en el tiempo series de sucesos, procesiones de seres que vienen de una cosa confusa que no se sabe bien si es fuerza o materia, pero se presenta como nebulosa disforme o como materia ya condensada, y se reparte en fragmentos, y constituye astros, y crea formas, no como quien se dispersa, sino más bien como quien se organiza. Y tras de la forma aparecen las cosas y las criaturas. Todo con un roce callado que va produciendo fulgores, estrellas, músicas y danzas. Se nos deslumbran los ojos, se nos embriagan los oídos, se nos deleita el corazón. La oreja hecha para los vientos y rumores de la tierra, recoge las vibraciones toscas; pero el ritmo infinito permea nuestras almas y las renueva de dicha y de gloria. La vida universal nos deja perplejos.*[19]

Universal life is bewildering because it exceeds the usual apparatus of detection—sensory perception and reasoning—that in the end are only capable of capturing "the rough vibrations" of the earthly world, but not "the infinite rhythm" of the primordial energy. The intuitive understanding of the path of creative force entails not only grasping the upheavals and purposes of this energy, but also facilitating the transcendent movement toward the Absolute of that which exists. This ultimately means that the human spirit has the responsibility of extending the uplifting movement of the energy: in the future the human soul "will extend the entire cosmos into the open sky" ("prolongará todo el cosmos hacia el cielo abierto").[20] This is, as we will see in the last section, the ultimate objective of the cosmic race.

In view of facilitating the transcendent movement, the philosopher must start from a partial and limited knowledge of things and then, in a supreme attempt at synthesis, coordinate the parts and reconstruct the unified harmony: "a true philosophy must start with the definition of God; but considering that today knowledge is disintegrated, it is necessary to begin by picking up the rosary beads and going over them carefully before proceeding to link them together" ("una verdadera filosofía tiene que comenzar con la definición de Dios; pero dado que el saber está hoy desintegrado, es menester comenzar por recoger la cuentas del rosario disperso y revisarlas cuidadosamente antes de proceder al engarce").[21] Thus, Vasconcelos divides his philosophical system into three main sections: metaphysics, ethics and aesthetics, each one providing the partial knowledge that is required to integrate them into an all-encompassing vision. Metaphysics studies the creative energy in its rational manifestations; that is, it investigates aspects of the world that can be known through the senses and intelligence. Ethics, on the other hand, deploys intuition or emotion to analyze energy as an action; namely, as a movement that shows a purpose or objective of spiritual ascent. Finally, through the lens of aesthetic contemplation or mystical revelation, aesthetics examines the moments in which energy auspiciously approaches its ultimate purpose. As a whole, metaphysics, ethics, and aesthetics successively replicate the path of the primordial energy through the physical domain, then the biological sphere, and eventually the spiritual realm.

In *La revulsión de la energía*, Vasconcelos provides an image to elucidate the distinctive natures of metaphysical understanding and aesthetical comprehension. He discusses the impact of cosmological forces, particularly sunlight, on Earth: the heat coming from the sun provokes diverse air currents and other weather phenomena, which in turn lay the groundwork for seasonal changes and their influence on living organisms. One may understand this cosmological process through the information provided by the senses and intelligence as modern science does. But Vasconcelos goes on to contend that in these same natural processes there is also a quality that can only be grasped with the help of an aesthetical lens: "In the periodic and perpetual change there is more than just renewal of biological processes, there is a beginning of beauty that only souls perceive" ("En el periódico y perpetuo cambio ya hay algo más que renovación de procesos biológicos, hay un comienzo de belleza que sólo perciben las almas").[22] While science might possess the tools for explaining the physical laws of air currents, aesthetical contemplation has the capability of grasping the "inherent laws" underlying the universe: "Man raises his mind and conforms to the winds, rolls in its eddies and also perceives something like the ultraphysical dynamics . . . that pulse in the cosmos" ("el hombre levanta la mente y se amolda a los vientos, rueda en sus giros y percibe también algo como la dinámica ultrafísica . . . que palpita en el cosmos").[23] In short, aesthetics provides insight into how the creative force functions and what its ultimate objectives are.

In *Tratado de metafísica*, Vasconcelos offers another example of the essential difference between a metaphysical approach and an ethical–aesthetical perspective by alluding to the experiments on the artificial creation of life that were being performed at the time. Even though he does not refer explicitly to Herrera's plasmogeny, it is highly probable that Vasconcelos was aware of his countryman's research in this area. Vasconcelos argues that so far, no scientist has been able to create an organism possessing functions such as growth, reproduction, and other characteristics of life, but he does not rule out that prospect. Even more, Vasconcelos asserts that it would not be surprising if life could be synthetically created: "Given that the affinity that exists between us and nature is so close, it would not be extraordinary if we were able to bind and unbind life" ("Dado que es tan estrecha la afinidad que existe entre nosotros y la naturaleza, nada de extraordinario tendría que llegásemos a atar y desatar la vida").[24] Human beings are already capable of controlling and transforming life through metaphysical means; that is, through the work of the senses and reasoning. But a truly transcendent endeavor, devoted only to an ethical–aesthetical approach, would be not merely to create and sustain life, but to *improve* it: "What the cell cannot do, transform energy beyond life and improve it, that is the work that is being tested in the laboratory of human consciousness; the most important of the internal tasks" ("Lo que no puede hacer la célula, transformar la energía más allá de la vida y mejorarla, eso es la labor que se ensaya en el laboratorio de la consciencia humana; la más importante de las labores internas").[25] In other words, only through intuition, contemplation, and revelation may humanity be able to pursue the goal of *overcoming* biological life and achieving the ascension of the primordial energy.

This veiled allusion to plasmogenic experiments also sheds light on the similarities and divergences between Vasconcelos's and Herrera's notions of universal life. Both intellectuals build on the idea that there is a universal flow of energy or matter that creates and underpins everything that exists. The diverse products of this universal flow—organic and inorganic matter, living organisms, space bodies, and so forth—are only differentiated by degree, not by qualitatively distinctive substance. However, while Herrera maintains that universal life shows no other objective than endless transformations and circulation, Vasconcelos assumes that there is a teleological flow of energy in the universe. Accordingly, Herrera suggests that the cosmic circulation of matter experiences continuous and contingent encounters, while Vasconcelos claims that the creative force undergoes successive phases, each more developed than the previous one, until reaching its final destination. Unlike Herrera's materialist and rationalist stance, Vasconcelos envisions through intuition that the final stage of the primordial energy will consist of a perfected spiritual realm. He thus presents human beings as the privileged instrument of universal life in the process of reaching the Absolute/ spirit. However, as we will see next, Vasconcelos's theorization regarding nonhuman ethics—which resonates with Herrera's own chemical ethics— puts pressure on the stability of his philosophical anthropocentrism.

NONHUMAN ETHICS

According to Vasconcelos, universal life displays an *ethical* behavior, since ethics pertains to everything that exists—not exclusively to human beings— as long as it deploys a deliberate action aimed at achieving uplifted forms of energy. For a reader of Western philosophical tradition, Vasconcelos's idea of nonhuman ethics constitutes a disconcerting, even foreign thought. As Guillermo Hurtado has put it recently, "The first thing that stands out about Vasconcelos's *Ética* is that it deals not only with humanity but with the entire universe" and accordingly "there is ethics in cells, plants, and the lower and higher animals" ("Lo primero que llama la atención de la *Ética* de Vasconcelos es que no sólo se ocupa de la humanidad sino del universo entero" y por lo tanto "hay ética en las células, las plantas, los animales inferiores y superiores").[26] Even though he recognizes nonhuman ethics as one of the most salient features of Vasconcelos's approach to ethics, Hurtado subsequently proceeds to dismiss it entirely focusing on the strictly human aspects of the theorization. He clearly states: "The section on the ethical hierarchy of animals stands out for its extravagance.... These occurrences are curious, even humorous—for those who do not love cats and dogs—but their proper place was not in the *Ética* but in a light magazine article. A conscientious editor would have deleted them" ("Destaca por su extravagancia la sección sobre la jerarquía ética de los animales. . . . Estas ocurrencias son curiosas, incluso simpáticas—para quienes no aman a los perros y gatos—pero su lugar no estaba en la *Ética* sino en un artículo de revista ligera. Un editor concienzudo las hubiera tachado").[27] Hurtado's unwillingness to engage with nonhuman ethics—and even his joking rejection—represent the standard reading that Vasconcelos's speculations have received in otherwise attentive overviews of his philosophical works.[28]

Rooted in unquestioned humanist assumptions, scholars of Mexican philosophy have generally been unwilling to take Vasconcelos's theoretical elucubrations on nonhuman ethics seriously and give them a more careful reading, which is what I aim to do in this chapter. My analysis goes along the general lines of David Dalton's posthuman reading of Vasconcelos's work, which foregrounds the nonhuman role of technology in ushering in an improved (post)humanity.[29] Similarly, I will analyze how plant life provides Vasconcelos with an ethical model for imagining the cosmic race utopia. By studying his conceptualization of the ethics of minerals, animals, and plants, I will also show that Vasconcelos simultaneously reinforces and contests the traditional hierarchy of beings that endows humanity with remarkable privileges over all other beings. As Susan Antebi maintains in her reading of *Ética*, "Although Vasconcelos does at times postulate the eradication of individuals or categories, at other moments he affirms essential equality between all beings."[30] This fundamental contradiction lies at the heart of Vasconcelos's engagement with nonhuman beings and even at the core of biocosmic thought at large. As we will see, even when Vasconcelos strives to subvert the established hierarchy of beings—by deeming plants the highest form of life, for example—he sustainably operates an anthropomorphization of those beings, suggesting that they paradoxically display a more human way of ethical behavior. Even so, I will show how the notion of botanical ethics unexpectedly opens up the possibility of disrupting anthropocentric frameworks.

In line with Herrera's theorization of the "life of metals," Vasconcelos maintains that one may find "ethical traces" ("sugestiones éticas") in the inorganic world when a mineral self-organizes and "collaborates, prepares and acts as a support for processes that transcend it" ("colabora, prepara, sirve de asiento a procesos que lo trascienden").[31] This ethics of minerals provides the material organization of every higher life form, but it remains mostly unacknowledged: "the ethics of minerals that is essential for life goes unnoticed" ("la ética mineral indispensable a la vida nos pasa inadvertida").[32] Yet the ethical behavior of minerals does not only lay the bedrock for biological life, it also displays specific features—such as harmonic proportions and color patterns, among others—that provide them a sense of identity or individuality: "In the purely mineral combinations that generate color, we distinguish a determination that obeys certain fixed rhythms and produces phenomena that are an imitation of personalization and individuality ... that is why we have always spoken of a character of steel and the hardness of diamond" ("En las combinaciones puramente minerales que engendran color distinguimos una determinación que obedece a ciertos ritmos fijos y produce fenómenos que son un remedo de la personalización y de la individualidad ... por eso se ha hablado siempre de carácter de acero y de firmeza de diamante").[33] In other words, the "chemical life of minerals" manifests attributes that, being grounded on material processes, suggest spiritual features such as tenacity and determination. This is why human beings are instinctively drawn to establish linguistic comparisons between their own realities and the behavior of minerals.

Even if minerals might display ethical behavior, the realm of ethics proper begins when energy undergoes an upheaval and creates cells and the biological sphere. Animals and plants already demonstrate—in the eyes of the philosopher truly in tune with the essential order of the world—movement in an upward direction, which of course becomes much more evident in human beings: "Life has known where it is going from its beginning as infusoria up to its completion in human consciousness" ("La vida sabe adónde va desde que comienza en los infusorios hasta que termina en la conciencia humana").[34] This movement toward the Absolute does not exclude times of retreat and descent that, in the case of the human being, give rise to what Vasconcelos calls evil and sins. Even so, the degree of elevation toward the spirit can constitute a fundamental kind of criterion according to which all life forms can be measured and hierarchized. It is a hierarchy that does not answer to properly biological, morphological, or evolutionary purposes, but to essential norms that have been discerned through intuition and revelation. Regarding the zoological sphere, unlike the evolutionary thesis that presented primates as animals closely akin to human beings, Vasconcelos considers other species as the most perfected and uplifted life forms.[35]

If it holds true that "animals, by their own activity, are not outside the plan of the Universe" ("los animales, por su actividad, no son ajenos al plan del Universo") it is also indubitable that certain animals seem to be more attuned to the universe's ultimate purpose.[36] Vasconcelos proposes a broad classification: "animals adaptable to our ethics" ("animales adaptables a nuestra ética") and animals "external to our purpose" ("ajenos a nuestro propósito") because they are either wild, harmful or indifferent to humans.[37] The first category is clearly superior, as these animals have contributed to human civilization, which in turn has functioned as the main vehicle of the creative force. But Vasconcelos further divides these ethical, superior animals into "subservient" (serviles) and "free" (libres), suggesting that there is an incompatibility between domestic animals such as dogs and cats and free animals such as birds and horses.[38] While both types serve or have served utilitarian purposes throughout human history, domestic animals are undignified companions that tend to foster human pettiness and divert the ascendant path of energy. In contrast, in addition to their functional uses, free ethical animals possess features—boldness, nobility, beauty, cleanliness—that showcase the higher faculties of the spirit. In particular, Vasconcelos devotes several pages throughout his work to the analysis and praise of horses, which in his opinion are "the perfect animal type" ("el tipo del animal perfecto") and "the cleanest, the noblest of all beasts, without excepting man" ("la más limpia, la más noble de todas las bestias, sin hacer excepción del hombre").[39]

In "Caballos: Velocidad"—an essay contained in *El pesimismo alegre* (1931)—Vasconcelos traces a cultural history of human-horse relationships since the times of ancient Asia, when for the first time "the alliance of the best of animals with the unsavory species of man was consummated. I wonder whether the horse has repented" ("se consumó la alianza del mejor de los animales con la especie turbia del hombre ¿Se habrá arrepentido el caballo?").[40] Rather than reinforcing a normative hierarchical relation between humans and *their* animals—objects of human possession and control—animal studies set out to investigate the various entanglements/alliances of human and nonhuman animals in which both agents demonstrate the power to shape their relationship. In consonance with this perspective, Vasconcelos recognizes the agency of nonhuman animals in their entanglement with humans, and even wonders if horses lament their historical connection to humanity. Vasconcelos goes on to trace some of the significant roles played by horses throughout human history: harnessing the physical force of horses, ancient human societies were able to travel long distances and wage war to expand their territory; horses appear in visual depictions alongside pharaohs and performers in Egypt; in Greek mythology horses populate the sky (Pegasus) and the ocean (Hippocamp); horses contributed to the conquest of the Americas and encouraged new social figures such as the Mexican "charro" and the American cowboy. In modern industrial societies, most of horses' functional uses—war, transportation, industry, agriculture—have been substituted by modern technology, making them expendable. The horse "seems a species that sees its mission accomplished and prepares to leave with the same noble elegance with which it has lived" ("parece una especie que ve cumplida su misión y se dispone a marcharse con la misma noble elegancia con que ha vivido").[41] But even if horses do not seem to have social relevance anymore, for Vasconcelos they still have plenty of ethical lessons to teach to humanity.

Traditionally, animals are considered to represent the purely biological part of human beings, and they often serve as a counterpoint to what is truly "human" in humans. In contrast, for Vasconcelos horses represent the best qualities that humanity must aspire to emulate: horses might be domesticated but they are always free; they selflessly cooperate with human intentions but never fully adopt conflictive human emotions; they do not waste energy in loving or hating other creatures; their reproduction is untainted by sexual lust and desire; they do not eat other animals and are utterly beautiful and clean. As Vasconcelos puts it, "you neither devoured a living species to live, nor did you fight with destructive zeal, nor did you participate in the homicidal madness of man. No beast equals you in nobility" ("Ni devoraste especie viva para vivir, ni luchaste en afán destructor, ni participaste de la locura homicida del hombre. Ninguna bestia se iguala contigo en nobleza").[42] Above all, horses represent for Vasconcelos unbound force and rapid velocity for traversing space with "a longing for distances and domination" ("un anhelo de distancias y de dominación").[43] The "appetite for horizons" ("apetito de horizontes") of horses makes them the perfect encarnation of the cosmic primordial energy, so much so that "The rhythm of horses marching, galloping, perhaps introduced the notion of force" ("El ritmo de los caballos en marcha, en galope, inició quizás la noción de fuerza").[44] Human beings have harnessed the force of horses in order to extend civilization throughout the planet, and in so doing have prolonged the path of the primordial energy further into the universe.

However, humans—a species riddled with appetites and so-called "aberrations" ("aberraciones")—have often betrayed the unbound and untainted force displayed by horses and must learn the ethical lessons of their nonhuman allies in order to better serve a cosmic function.[45] For example, once horses became useless in modern societies, humans have tended to disregard them as insignificant beings, or even worse, have forced them into undignified spectacles such as horse racing. But horses remain characteristically unfazed: "And the horse, always indifferent, with no illusions about man, continues to drink water and eat hay; he continues with his sinless eyes, looking with embarrassment at the strange beast that climbs onto his back, pulls him by the bridle, makes him run or leaves him forgotten. The horse, meanwhile, lives by its instincts and its love of the air and the vastness" ("Y el caballo, siempre indiferente, sin ilusiones acerca del hombre, sigue bebiendo su agua y comiendo su paja; sigue con sus ojos sin pecado, mirando con azoro a la extraña bestia que ya se le sube al lomo, ya lo hala de la brida, ya lo pone a correr o lo deja olvidado. El caballo, en tanto, vive sus instintos y su amor de los aires y de la inmensidad").[46] In this quote, Vasconcelos adopts the horse's perspective to present the world as purportedly experienced by the animal. If animal studies have cautiously explored the question of the (im)possibility of truly grasping the animal point of view without incorporating human frameworks, Vasconcelos seems to maintain that a philosopher in tune with the essential order of the universe can and should understand the experience of every being on Earth.[47] From the point of view of the horse, humans are "strange beasts" that do not seem to embrace the fundamental thrust of the energy toward the open universe, "its love of the air and the vastness."

While Vasconcelos greatly admires horses as ethical models, his opinion about animals in general is mostly unfavorable. It can then be said that animals, due to their biological closeness to the human being, represent the component that corrupts and pushes humanity back to its immediately inferior stage or prevents it from rising spiritually. Horses—and to a lesser extent birds—are the exception in this general framework. In contrast, according to Vasconcelos's ethical scale, plants occupy a privileged place that even the animal kingdom must aspire to imitate: "Both in their habits and in their relationship with things human, plants demonstrate an incomparably cleaner and more elevated behavior than zoology" ("Tanto en sus hábitos como en sus relaciones con lo humano, la planta ofrece un sistema de conducta incomparablemente más limpio y elevado que la zoología").[48] Plant life not only adapts itself to the utilitarian (agriculture, medicine) and aesthetic (object of representation and decoration) purposes of human beings, but more importantly plants constitute in themselves a way of life in which one can easily appreciate the will to reunite with the Absolute. While the animal kingdom "gives us the impression of a failed order, dirty and inept . . . how easily we accept the immaculate divinity of leaves, vines, trunks, meadows in bloom and luxuriant woodlands" ("nos da la impresión de un orden malogrado, sucio e inepto . . . con qué facilidad aceptamos la divinidad inmaculada de los follajes, las lianas, los troncos, los prados floridos y los montes frondosos").[49] Even the human being tends to betray his very essence as an instrument of ascension of the spirit, but vegetal life usually shows an irreproachable ethical integrity: "Plants never backslide ethically. The spirit in them moves slowly, but it does not go backward" ("La planta nunca camina hacia atrás éticamente. El espíritu va en ella lenta, pero no retrocede").⁵⁰ Hence, Vasconcelos sets out to "base (human) morality on the 'behavior,' in the system of life of plants in order to contrast this botanical morality with the zoological ethics that it is wished to impose upon us" ("fundar la moral [humana], en el behaviour, en el sistema de vida de las plantas, a fin de oponer esa moral botánica a las éticas zoológicas que nos quisieron imponer").⁵¹

Despite being—or precisely because they are—very distinct life forms from the human being, plants contain a common substrate that essentially links them to humanity: plant life is the factor within human biology that drives it to definitively transcend the biological sphere. In this way, Vasconcelos complicates yet does not ultimately disrupt the biological hierarchy that has been repeated since its formulation in classical antiquity. According to Aristotle, plants show the simplest traits of life (nutrition, growth, reproduction), while animals add perception and movement to the vegetative functions, and humans culminate the hierarchy due to their use of reasoning. Plants display, then, the minimum and indispensable functions to qualify as living beings. For Vasconcelos, plants also represent the least sophisticated form of life, but according to him, this does not imply a negative or pejorative assessment. On the contrary, precisely because they are elementary life forms, plants clearly show the essential potency of life—understood as an action oriented toward the Absolute—without the "drawbacks" and "sins" observable in the supposedly higher beings. Vegetal life is "divine" and aesthetically pleasing because it coincides fully with the transcendental impulse that originated that which exists. At the same time, however, plants are unable to carry out the ultimate purpose of energy: the last upheaval that will culminate in the creation of a spiritual and supernatural order. Such a purpose may only be undertaken by a particular being, the human being, who paradoxically must remain faithful to their vegetal substrate if they desire to avoid losing sight of the transcendental purpose of energy. Thus, the traditional biological hierarchy is preserved in Vasconcelos's thought, although it is clearly transformed according to his particular conception of life.

As can be seen, Vasconcelos admires plants and horses for the same general reasons. However, as we will discuss next, his ethical proposal specifically seeks inspiration in the life of plants so as to define a model of behavior that distances itself from the so-called "survival of the fittest;" "seeking in plants less brutal, *more human* standards than those derived from zoology by evolutionism" ("buscar en las plantas normas menos brutales, *más humanas* que las derivadas de la zoología por el evolucionismo").⁵² It is, without a doubt, a process of anthropomorphizing plant life: the human traits and qualities that are considered desirable are projected onto vegetal life, so that plants hold the ideal that was sought in them beforehand. Nevertheless, it is necessary to emphasize that Vasconcelos tries to base his ideas on plant behavior on knowledge derived from the botanical investigations of his time, many of which are still considered pioneering in the field.[53] Although Vasconcelos rejected the reduction of all knowledge to the knowledge extracted from the sciences, which has led some critics to talk about the antiscientific nature of his thought, the truth is that he believed that philosophy should draw findings from science in order to later face metaphysical questions.[54] Vasconcelos was chiefly reacting against the positivist dogmas of his time without eschewing science as a whole: "(he) frequently and paradoxically turned to scientific discourse to buoy his antipositivist ideas."[55]

THE HUMAN-PLANT HYBRIDS OF FRIDA KAHLO

Before turning to Vasconcelos's theorization of botanical ethics, it is productive to delve into Frida Kahlo's engagement with plant life in the context of her cosmological ideas. Both Vasconcelos and Kahlo showed a distinctive concern with the life of plants and its interconnection with the Earth, human beings, and the cosmos. Kahlo's rise to global fame—which came a few decades after her death in 1954—was underpinned by an accentuated emphasis on her life: her chronic health problems, frenetic marriage to Diego Rivera, her personal lifestyle and sense of fashion, and so on. It is not surprising that Kahlo's work has predominantly been interpreted in the context of her biography, making clear connections between the dramatic events of her life and her visual work.[56] This has led to a critical overemphasis on Kahlo's self-portraits that evoke personal stories at the expense of other areas of her work such as portraits and her large production of still-life paintings. Overall, there has been a close examination of fundamental aspects of her work, such as her incorporation of folk and Prehispanic motifs, her connection with surrealism, and her complex exploration of gender and race in the context of the postrevolutionary discourse of *mestizaje*, among others.[57] Much less scholarly attention has focused on Kahlo's engagement with botanical or cosmological themes, and when scholars have explored these issues, they have frequently connected them with postrevolutionary notions of race, gender, and identity, as we will see next.

Retrato de Luther Burbank (1931) is a superb illustration of how shifting the focus from the predominant emphasis on race or gender to a biocosmist reading could open up new pathways for engaging with Kahlo's work. Likely because it is a portrait and not a self-portrait, this work has received

112   BIOCOSMISM

FIGURE 3.1. Frida Kahlo's *Retrato de Luther Burbank*. Provided by Schalkwijk / Art Resource, NY. © 2024 Banco de México Diego Rivera Frida Kahlo Museums Trust, Mexico, D.F. / Artists Rights Society (ARS), New York.

comparatively less scholarly and public attention than Kahlo's famous paintings, which typically include a depiction of her now-iconic face and *Tehuana* dress. In contrast, this painting portrays botanist and horticulturist Luther Burbank (1849–1926), who became renowned internationally during the first decades of the twentieth century for creating varieties of hybrid trees, flowers, fruits, and other plants. In 1930, a few years after Burbank's death, Kahlo and Rivera—who would eventually create themselves a famed garden at their residence in Coyoacán—visited the botanist's house and garden in California and learned about his life and work.[58] Famously, Burbank had instructed that he be buried directly under a tree so that his decaying organic matter would nourish the plant. Both Kahlo and Rivera would go on to paint separate portraits of Burbank in 1931, which speaks of the interest and admiration they surely felt for him after visiting his garden. Yet whereas Rivera's image of Burbank grafting a plant in *Alegoría de California* (1931) at the Pacific Stock Exchange in San Francisco serves the function of emphasizing the technological expertise and natural resources of California,

Kahlo's portrait is an exploration of a "cosmic framework of life and death" that relies on fantastical or surrealist imagery of Burbank's burial.[59]

In *Retrato de Luther Burbank*, we see the renowned horticulturist portrayed as a human-plant hybrid: his human-like upper body is holding a philodendron—a species native to Mexico—while the lower part of his legs resembles a tree trunk whose roots reach and penetrate a realistically portrayed dead body underground. As Nancy Deffebach has suggested, this portrait is alluding not only to Burbank's burial and his profession of plant hybridization, but probably also to his philosophical suggestions contained in *The Training of the Human Plant* (1907), which draw an analogy between plants and people on the basis of their shared dependence on environmental circumstances and hereditary factors.[60] In contrast to the social Darwinist idea of "survival of the fittest" which was based on zoology, Burbank—and, as we will see, Vasconcelos too—believed that botanical hybridization offered a better framework for understanding the "cultivation" of desired traits of human populations.[61] While the extent to which Kahlo was familiar with Burbank's theories is unknown, it is safe to say that she at least was aware of their general outlook and main propositions. In this line, Lucretia Hoover Giese has put forward the idea that Kahlo was captivated by Burbank's hybridization theories because they enabled her to symbolize her own mixed Mexican and European identity and, more generally, the postrevolutionary construction of *mestizaje* as the basis of *mexicanidad*.[62] Additionally, by depicting a renowned American botanist holding a plant indigenous to Mexico, Kahlo was probably addressing her mixed emotions toward her own experiences living in the United States as a proud Mexican, and the prospect of Pan-American cooperation.[63]

While this is a productive reading of *Retrato de Luther Burbank*, I would like to propose an interpretation that, instead of relying on the predominant theme of national identity, draws attention to Kahlo's broader conception of the vitality of the cosmos. This portrait explores the "interrelatedness of all life," particularly of plant and human life, suggesting that these two disparate forms of life are intimately connected and interdependent.[64] Whereas human beings depend on the oxygen, food, and other products provided by vegetal life, certain varieties of plants owe their existence and proliferation to horticulturists such as Burbank. More broadly, by depicting a continuity between human and vegetal bodies, this portrait suggests that human beings and plants are two specific, essentially comparable variants of universal life. While this suggestion destabilizes the traditional biological hierarchy, the portrait similarly puts pressure on the division between the living and the nonliving, or the organic and inorganic spheres, by depicting an underlying continuity between the two realms. Thus, this work is not only a representation of "death nurturing life," but also a suggestion that there is no essential distinction between the living and the inert: the two are particular manifestations of the unitary, universal flow of energy and matter, as both Vasconcelos and Herrera argued.[65]

Furthermore, *Retrato de Luther Burbank* provides a clear illustration of the notion of plants defined as "cosmic mediators," that is, agents that because of their own physiology are bound to establish close relationships with the soil and the air, the Earth (the geological) and the sun (the cosmological).[66] Located at the center of the work, the image of the human-plant hybrid is in effect interconnecting the three zones of the landscape: cosmos, earth, and underground, clearly demarcated by distinct colors. The blue cosmos contains a few white and gray clouds whose shapes bear a significant resemblance to the large plant leaves to the left of the horticulturist. The light brown earth holds two lush trees with hanging fruits, seemingly an allusion to Burbank's work of producing high-yielding varieties of plants. Lastly, the dark brown underground displays small root systems adjacent to the human-plant hybrid's roots drawing nourishment from the decaying corpse. Thus, Kahlo's painting theorizes hybridization not only as a *racial* process—as Lucretia Hoover Giese argues—but more fundamentally as a *cosmic* process: it is a mixture taking place not only between racially diverse humans, but also between humanity and plants, between the cosmos and the Earth, the organic and the inorganic, establishing a far-reaching, endless process. As in Rivera's *El hombre controlador del universo*, discussed in Chapter 2, Kahlo's protagonist is located "at the crossroads" of universal life: his fundamental task consists of fostering the proliferation of the cosmic hybridization process.

Depictions of human–plant hybrids appear in several other works produced by Kahlo in the 1930s and 1940s.[67] One of the most significant in terms of its biocosmic overtones is *El abrazo de amor del Universo, la Tierra (México), yo, Diego y el señor Xólotl* (1949), which similarly portrays entanglement and indistinction between the biological (plants, animals, humans), the geological, and the cosmological.[68] Whereas *Retrato de Luther Burbank* employs a layered landscape composition—including a cross-section of the Earth—to address the interconnection of the universe, *El abrazo de amor* incorporates a structure of concentric circles which allegorically encompasses the diverse realms of the whole cosmos. In the outermost circle we see an anthropomorphized universe against a dark and bright background containing clouds and celestial bodies. The enormous arms of the universe—

FIGURE 3.2. Frida Kahlo's *El abrazo de amor del Universo, la Tierra (México), yo, Diego y el señor Xólotl.* Photo courtesy of Erich Lessing / Art Resource, NY. Frida Kahlo, © 2024 Banco de México Diego Rivera Frida Kahlo Museums Trust, Mexico, D.F. / Artists Rights Society (ARS), New York

one dark, the other bright—are holding lush vegetation and a sleeping dog, as well as a green human-plant Earth whose lactating breast has a large crack and is nurturing a tree. The Earth is embracing a weeping Kahlo who in turn cradles a nude, child-like Diego Rivera in the innermost circle. Overall, while it includes marks of conflict introduced by the cracking breast and Kahlo's tears, the painting seems to suggest that the universe ultimately neutralizes these binary relations in a harmonious totality, just as it thrives against the dark and bright background.[69] In the end, this work suggests an interdependent, loving relationship between each realm of the universe, which simultaneously embraces a dependent body and is being embraced by a higher being.

El abrazo de amor is perhaps the best expression of what Hayden Herrera calls Kahlo's "vitalistic form of pantheism," which is closely related to biocosmic thought.[70] Like the other intellectuals studied in this book, Kahlo conceived of the universe as a unitary, immanent domain which comprises a plurality of processes and beings that, despite their differences and specificities, share an essential sameness. This cosmic uniformity is granted by the fact that every being and realm, both in the organic and inorganic spheres, is animated by a sort of vitality that endows them with sentience, movement, and expressiveness. Throughout her work Kahlo—like Vasconcelos—

paid special attention and consideration to the expressive life of vegetal beings (trees, fruits, flowers, plants), which feature prominently in *El abrazo de amor* and other self-portraits, as well as in her still-life paintings. As I showed earlier, Kahlo's approach to plant life often highlighted the identity between plants and humans, and in some cases was associated with broader speculations about the nature of the cosmos. Kahlo, in line with Alfonso L. Herrera's notion of biocosmic solidarity, believed that there was an intrinsic affinity and sympathy between all members of universal life, despite—or precisely because of—the all-pervasive binaries and antagonisms in nature. By the 1940s Kahlo's works recurrently "portray her integrated into a massive, undifferentiated Web of being," as Gannnit Ankori suggests.[71]

Some of these biocosmic ideas were condensed by Kahlo in a stream of consciousness written in her diary in the 1940s. This text provides us a glimpse of the general lines of thinking that presumably underpinned the creation of *El abrazo de amor*. She starts by describing "la vida" as "el viaje atómico y general:" a totality in which human beings are forcefully embedded: "No one is more than a function—or part of a total function. Life goes by, and sets paths, which are not traveled in vain. But no one can stop 'freely' to play by the wayside, because he will delay or upset the general atomic journey" ("Nadie es más que un funcionamiento—o parte de una función total. La vida pasa, y da caminos, que no se recorren vanamente. Pero nadie puede detenerse 'libremente' a jugar en el sendero, porque retrasa o transtorna el viaje atómico y general").[72] Human beings are not understood as agents with free will who are able to stand above and control universal life, but rather as fields uncannily traversed by the universal flow of atoms: "Fear of *everything*—fear of knowing that we are no more than *vectors direction* construction and destruction to be *alive*" ("Miedo a *todo*—miedo a saber que no somos otra cosa que *vectores dirección* construcción y destrucción para ser *vivos*").[73] Thus, rather than being hierarchically separated from the universe, the life of human beings is comprised of the same basic matter as the rest of the beings in the cosmos and it is fundamentally akin to them. For Kahlo, human existence entails the sudden realization that there is something of everything in everything; humanity *is* plants, rocks, stars, microbes, and so on:

> taking part in the complex *current* not knowing that *we* are heading toward *ourselves*, through millions of stone beings—of bird beings—of star beings— of microbe beings—of fountains beings toward ourselves—a variety of the *one* incapable of escaping to the two—the three—to the usual—to return to the *one*. Yet not to the *sum* (sometimes called *god*—sometimes *freedom*,

sometimes *love—no—we are hatred—love—mother—child—plant—earth—light—ray—as usual—world bringer of worlds—universes and universe cells.*

participar en la corriente compleja de no saber que nos dirijimos a nosotros mismos, a través de millones de seres piedras—de seres aves—de seres astros—de seres microbios—de seres fuentes a nosotros mismos—variedad del uno incapacidad de escapar al dos—al tres al etc de siempre—para regresar al uno. Pero no a la suma (llamada a veces dios—a veces libertad a veces amor—no—somos odio—amor—madre—hijo—planta—tierra—luz—rayo—etc—de siempre—mundo dador de mundos—universos y células universos.[74]

Kahlo's engagement with the notion of universal life allows us to read her work productively as part of the intellectual constellation of biocosmism. Particularly, as I will show next, her conception of plant life in connection with the process of cosmic hybridization is in remarkable agreement with Vasconcelos's theorization of plant consciousness, which similarly builds on the philosophical implications of the close interaction between plants and their natural surroundings. In addition to plant consciousness, Vasconcelos's idea of plant collectivity serves as the basis of his formulation of botanical ethics and of the utopian program of *La raza cósmica*. In the end, in parallel with Kahlo's exploration of human–plant hybrids, Vasconcelos's utopia of the Cosmic Race speculates about the ultimate consequences for humanity of adopting botanical ethics: *being* like plants in their specific form of ethical and collective behavior will have transcendental repercussions for the course of the whole universe.

PLANT CONSCIOUSNESS

In Vasconcelos's historical moment and still in our days, talking about plants' behavior could seem far-fetched for many biologists and lay people. Since the habits of most plants remain unnoticeable to ordinary human perception—their movements, for example, mostly occur with a speed that cannot be perceived with the naked eye—the notion that plants are static, insensitive and fundamentally passive beings has been widely propagated.[75] In order to counteract this idea, Vasconcelos had to show, first, that plant life has the capacity to capture environmental stimuli and act based on the information obtained in that process. Vasconcelos turned to the research of Sir J. Chandra Bose, who—as we saw in the last chapter—was also essential for Herrera's notion of chemical ethics. In the early twentieth century, Bose not only studied the responsivity of metals, but also launched an analysis of vegetal life, which puts him as a founding figure of the field of study of plant behavior.[76] Bose studied the species Mimosa pudica—well known for its sudden movements—and discovered through mechanical devices of his own invention that, when these plants receive a stimulus as minimal as touch, they emit an electrical signal that eventually generates the movement of leaves. Bose extended the conclusions of these studies to plant life in general: although they do not have a brain or a nervous system similar to animals, plants do use a complex "nervous mechanism" or electrical conduction system through which they react to sensitive stimuli provided by their environment.

According to Vasconcelos, these botanical investigations had notable philosophical implications: "Bose shows that it is necessary to slightly rectify what is often claimed regarding the blindness and insensitivity of matter, even when it is understood that in our own conception ... the differences of nature and its classifications are a matter of degree and not radically primordial" ("Demuestra Bose que es necesario rectificar un tanto lo que se afirma a menudo de la ceguera e insensibilidad de la materia, aun cuando ya se entiende que en nuestra propia concepción ... las diferencias de naturaleza y sus calificativos son cuestión de grado y no de raíz").[77] In other words, plants—apparently "insensitive" and "blind" lifeforms—suddenly appear as beings moved by the same essential energy that variously animates all that exists, even inorganic matter. In a more specific sense, Bose's experiments show that plant movements are the result of "a true nervous action as shown by the fact that they are affected by the same agents that affect the nerve: heat, cold, narcotics" ("una verdadera acción nerviosa según lo prueba el hecho de que la afectan los mismos agentes que afectan al nervio: el calor, el frío, los narcóticos").[78] Vasconcelos emphasizes that this "nervous action" is the product of a vital mechanism that should not be considered wholly reducible to the stimuli of the outside world; on the contrary, "the purpose of the changes in movement does not strictly correspond to the external stimuli, but rather to certain internal factors" ("el sentido de los cambios del movimiento no corresponde rigurosamente a los estímulos externos, sino a determinados factores internos").[79] In other words, plant sensitivity implies that there is "an original primary motor impulse arising from the cell" ("un impulso motor primario original, nacido de la célula") that reacts creatively to confront the challenges of the environment.[80]

In fact, the particular consciousness of plants arises precisely from their way of establishing a close relationship with the environment they inhabit.

Vasconcelos suggests that plants found their form of life through their need to constantly "feel" and "be aware" of the sensitive data they receive from the outside world. The fact that they are "fixed" on the Earth does not represent a disadvantage in comparison with the extreme mobility of animals, but rather a circumstance that forces them to develop an even closer interdependence with the environment:

> Plants are fixed and yet in communication with all the corners of the earth and sky; in greater communication with the earth and sky than moles and rabbits; even though rabbits are like us in that they walk to and from, and they have a special organ for digestive functions and a nervous system, eyes, ears. Who would dare deny that a palm tree is more aware if awareness is understood as connecting life itself with the breath of the elements and with the restlessness of souls?

> *Las plantas están fijas y sin embargo en comunicación con los rincones todos de la tierra y el cielo; mucho más en comunicación con la tierra y el cielo que los topos y los conejos; aunque los conejos andan como nosotros de aquí para alla, y tienen órgano especial para funciones digestivas y sistema nervioso y ojos y oídos, ¿quién osará negar que tiene más conciencia una palmera, si por conciencia se entiende ligar la propia vida con el hálito de los elementos y con la inquietud de las almas?*[81]

The apparent simplicity of plants—they do not have "special organs" that respond hierarchically to an organizing center—is revealed as a vital strategy for receiving and interpreting information with all parts of their body, from the roots that are connected to the earth to the leaves that are in permanent contact with the surrounding light and air.[82]

In this way, Vasconcelos offers a definition of consciousness— "connecting life itself with the breath of the elements and with the restlessness of souls"—that is proper to vegetal life and establishes a clear opposition to the idea of consciousness as a privileged product of the last "revulsion," or upheaval of energy that brought humanity into being. We are dealing, ultimately, with a tension between two notions of "nature": two socially constructed images of what the natural world is and what relationships it establishes with humanity. On the one hand, the idea of consciousness as an exclusive product of the human being implies an essential separation between Society and Nature, since the latter is conceived as an external and independent environment that opposes and challenges the human will. The perfect image of this conception of nature is the social construction of the "wilderness": the sublime landscapes—waterfalls, high mountains, canyons, etc.—that usually generate feelings of astonishment and religious contemplation.[83] It is also worth noting that the construction of the natural world as totally external is a precondition of the process of capitalist accumulation, since the latter is based on the assumption that organic and inorganic elements are resources waiting to be exploited by the human being.[84] The Society/Nature binary presumes, in short, that human beings have an ontological, political, economic privilege, which allows them to appropriate— up to the limit of devastation, if necessary—all natural resources in an instrumental way. This is why, in Vasconcelos's thought, the "transcendental plan" ("plan transcendental") of energy decrees that the human being will control, dominate and, finally, transcend the natural or biological sphere in search of spiritual realities.

At the same time, in Vasconcelos's philosophy this idea of consciousness coexists with another that calls into question the Society/Nature binary and destabilizes the power of the human being over the natural world. The notion of plant consciousness requires, at the outset, to think of nature as an interconnection of organic and inorganic elements in which the human being would occupy a position similar to any other entity. In recent decades, various concepts have been proposed—assemblage (Deleuze and Guattari), web of life (Moore), mesh (Morton), among others—to understand this interdependent network of relationships that neutralizes the clear divisions between the human and the nonhuman. Vasconcelos seems to approach this kind of formulation when he speaks of nature as a "vast laboratory" ("vasto laboratorio") that diversifies energy and creates different types of consciousness that depend on each other: "new observations concur in their suggestion that plants and animals act as tests in great numbers and types, through which nature releases its creative energy to constantly seek a broader zone, greater reach and, in short, consciousness through varying combinations. Man represents the greatest success of this vast laboratory; however, all beings support and build the environment for each other" ("nuevas observaciones concurren para indicar que las plantas y los animales son a manera de ensayos en multitud de número y de géneros, por los cuales la naturaleza da salida a su energía creadora, para buscar constantemente, en combinaciones cambiantes, mayor amplitud de zona, mayor grado de extensión y en suma, de conciencia. El hombre representa el mayor de los éxitos de este vasto laboratorio; pero todos los seres, unos a otros, nos servimos de sostén, nos construimos el ambiente").[85] The human being still has a privileged position in this framework, but the idea of an interdependence between beings suggests a path that tends to topple the perceived anthropocentrism of Vasconcelos's philosophical system.

Although Vasconcelos does not extract the environmental implications of these suggestions, the idea of plant consciousness as the basis of botanical ethics implies that the human being would have an ecological approach inspired by the close and "armoniosa" relationship of plants with the natural world: "We cannot even imagine how much contemporary social ethics derived from analogies with the beast would gain if it were to reflect on the harmonious action of the fibrovascular system (of plants) in relation to the elements. One feels justified to talk about a botanical soul when pondering about the emigration of pollen in the winds" ("No podemos imaginar cuánto ganaría la ética social contemporánea, derivada de analogías con la bestia, si reflexionase en la acción armoniosa que cumple el aparato fibrovascular (de las plantas) en relación con los elementos. Se siente justificado hablar de un alma botánica cuando se piensa en las emigraciones del polen por los vientos").[86] Vasconcelos seems to suggest that vegetal life stages a way of life that does not strive to establish a sovereign and dominant relationship over nature, but presents itself as a junction that extends the overlapping of all points within the network of relationships. The image of "the emigration of pollen in the winds" provides an appropriate point of departure for visualizing the far-reaching and unexpected interactions that maintain the web of life.

The fact that Vasconcelos highlights "the harmonious action of the fibrovascular system in relation to the elements" suggests that plants carry out a properly cosmic function, as Kahlo also expressed in her work. Vascular plants are characterized by vascular tissue that transport water, minerals, and nutrients throughout the plant. In order to do this, vascular plants must support themselves on two structural components: roots that absorb water and nutrients from the soil and stems that connect the roots to the leaves and elevate the foliage high above the ground to receive the required sunlight for photosynthesis. In this sense, vascular plants serve as "cosmic mediators," since they put into communication radically different environments: the soil and the air, the Earth and the sun.[87] Plants also interconnect and sustain all other living organisms, including human beings, due to their ability to produce oxygen, maintain the atmosphere necessary for life, and store chemical energy that fuels other life forms when ingested. The life of plants is by definition a *cosmic* process defined by the mutual interpenetration between disparate spheres: the organic and the inorganic, the Earth and the sun. Thus, Vasconcelos's project of theorizing botanical ethics is rooted in the cosmic function of plants: human beings must wholeheartedly embrace the cosmic behavior of plants.

On a philosophical level, the idea that vegetal life depends entirely on the elements (water, earth, air, light) introduces an "image of thought"—that is, an image of "what it means to think"—that challenges Vasconcelos's own philosophical model.[88] He uses trees and botany to explain the assumptions that underpin his various schemas: "For the philosopher, the tree's trunks and branches, the fragrance of the sap and the song of the leaves in the wind are variations of a wonder that must be penetrated beyond the superficial appearance, beyond the exterior that is translated in shapes, to the invisible core of the seed where the creative impulse is forged" ("Para el filósofo, el tronco y las ramas del árbol, la fragancia de las savias y el canto de las hojas al viento, son variantes de un prodigio que es menester penetrar más allá de la apariencia que nos da superficies, más allá de lo exterior que se traduce en formas, hasta el seno invisible de la semilla donde se forja el impulso creador").[89] Reproducing the founding dichotomies of Western metaphysics, Vasconcelos makes it clear that philosophy must think about essences (not accidents such as appearance and form) and it must grapple with the creative power—not limit itself to acts already consummated—in the pursuit of "penetrating" the true "wonder" of all that exists. In other words, the philosopher's goal is to get to know "the seed," the *arkhé* as the primordial energy and organizing principle of everything that exists.

For Vasconcelos, "thinking" means understanding the original principle that acts as a self-sufficient and self-determined cause of reality. This essential principle ultimately ensures that all beings remain identical to themselves (identity) and have a rational explanation that justifies their existence (principle of sufficient reason). In opposition to such an identitarian and principial model of thought that has predominated in Western metaphysics, a new possibility of understanding the act of thinking based on the "radical dependence" of plants on the environment emerges, which implies from the outset that being no longer finds a necessary justification in its own essence and identity.[90] In the same vein, Coccia proposes the notion of "metaphysics of mixture" to understand how plants—and the world as a whole—depend on the "mutual compenetration between subject and environment, body and space, life and medium."[91] Similarly, Marder poses the idea of the "heteronomy" (the law of the other) of plants: the notion that plant life finds its ontological foundation always outside of itself, that is, in the changing and immanent relationship they establish with the others.⁹² In this way, the idea of plant consciousness destabilizes the philosophical model of "the seed" and open up space for a new understanding of thinking.

Plants allow for a conception of being as a contingent and unfounded heterogeneity that is radically exposed to the Other (the alterity outside and inside the Self). It is, therefore, the possibility of a post-foundational and non-identitarian thought that, since the second half of the twentieth century, has found diverse elaborations inspired precisely by vegetal life, such as the concepts of "dissemination" (Derrida) and "rhizome" (Deleuze and Guattari). Both of these concepts put at stake the idea that being is difference: an immanent multiplicity of differences that allow for, and at the same time, continually destabilize the identities and oppositions that are built upon it. Of course, Vasconcelos does not come closer to these elucubrations that are seemingly incompatible with the spiritualizing thrust of his philosophy. But, even so, the notion of plant consciousness surprisingly unsettles the building of his philosophical system and points to the possibility of an equally an-archic (without *arkhé*) and ecological thought. Furthermore, plant consciousness allows Vasconcelos to suggest the properly cosmic function of plants and life in general: vegetal life becomes, then, the epitome of universal life.

PLANT COLLECTIVITY

In addition to plant consciousness, Vasconcelos highlights the collective behavior of plants as one of the theoretical bases of botanical ethics. The idea that plants, particularly trees, act as collective beings was recurrent in botanical writings from the eighteenth and nineteenth centuries, by authors such as Goethe and Darwin, among others.⁹³ Vasconcelos, specifically, builds on the work of Jean Henri Fabre (1823–1915), a French naturalist who is best known as one of the pioneers in the study of insects (entomology). While many entomologists derived their knowledge from the analysis of inert insects, Fabre devoted himself to the direct observation of what he called "the world of insects," that is, the behavior and habits of these animals as they constantly interact with their immediate surroundings. This perspective, which is more attentive to environmental relations than to the morphological characteristics of living beings, guided Fabre's research on the particular form of life of plants. His text *The Wonder Book of Plant Life*, originally *La plante: Leçons à mon fils sur la botanique* (1876), served as the basis for Vasconcelos's philosophical argument, which closely follows and sometimes paraphrases the observations of the French naturalist.

Fabre begins his work with a discussion on the ways in which each plant or tree consists of a group of individuals that extends over time, both in the past and in the future: "Each spring" states Vasconcelos in drawing from the French naturalist, "a new generation appears; the trunk and branches represent generations of the past; but they are not dead: they support the new and future generations" ("Cada primavera ve aparecer una nueva generación; el tronco y las ramas representan generaciones del pasado; pero no están muertas: sirven de soporte a las nuevas y a las futuras generaciones").[94] Each bud is potentially an individual that will grow with the collective support of the plant and will become a branch that will in turn generate other buds in the future. Thus, Vasconcelos concludes that the plant "is not a single being, but rather 'a collective being,' an association of individuals, all of whom are related, intimately united and working together for their common prosperity" ("no es un ser simple, sino 'un ser colectivo,' una asociación de individuos, todos relacionados, íntimamente unidos y trabajando en conjunto para la prosperidad común").[95] Plant behavior is, then, eminently collaborative in nature, both among the individuals that make up each plant and among the various plant species. This behavior, according to Vasconcelos, constitutes an essential difference with the Darwinian logic of "survival of the fittest" that underpins animal habits.[96] Although parasitic or carnivorous plants that feed on other species in order to survive exist, "it is more common to find plants that, even when living off another plant, do not destroy it, but rather find support in it and in turn produce a sort of robe that embellishes and protects" ("más frecuente es el caso de plantas que, aun viviendo de otra, no la destruyen, sino que se apoyan en ella y le fabrican una suerte de vestidura que embellece y protege").[97]

Vasconcelos argues that a more careful consideration of the form of life of plants would surely lead to an ethics removed from the struggle for existence and inspired by the "plant's architectural plan in the jungle" ("plan arquitectónico de la planta en la selva") instead.[98] For Vasconcelos, the jungle—not other ecosystems such as the forest or the desert—represents the perfect model that illustrates the collaborative behavior of vegetal life. Every plant cooperates differently in the jungle to build a self-sustaining and lasting environment. Even the species considered to be parasitic contribute to this general purpose: "Climbing plants, which use their spurs to steady themselves on trunks and rise, build the jungle's protective tangle" ("Las trepadoras, que usan su espuela para afirmarse en los troncos y ascender, construyen la maraña protectora de la selva").[99] Thus, the jungle implements the inherently deliberate and constructive activity of vegetal life, which "persists, indifferent to the distress of the destiny that wavers, unaware of the pain and responsibility, sure of its task; weaving its vast plan, repressing the clamor of its splendid triumph" ("perdura, indiferente a la angustia de los destinos que vacilan, ignorante del dolor y de la responsabilidad, segura de su tarea; tejiendo su vasto plan, reprimiendo el clamor de su triunfo espléndido").[100] The jungle embodies the notion of a community whose "destiny" does not "waver" because it acts according to a "vast plan" or "task." In other words, it is a community founded on a stable essence or identity, an unwavering "destiny" that establishes an original principle and ultimate purpose (*arkhé*).

The individual within this collective must conduct himself or herself according to a distribution of roles and forces: the "vast plan" or "task" that is necessarily determined by the primordial principle. As a result, this model of community entails a complete lack of some sort of social conflict or antagonism that would have arisen from disagreement with the established distribution of roles: all society members "are related, intimately united and working together for their common prosperity" ("todos relacionados, íntimamente unidos y trabajando en conjunto para la prosperidad común"). The "splendid triumph" of this collective is a self-sufficient existence, always identical to itself, in which the new generations—the leaves that are born from the trunks and branches of the tree—function as a mere extension of the same shared purposes. Ultimately, the heterogeneity that is inherent in the condition of being-in-common is subsumed within a homogeneous totality that does not leave room for freedom and self-determination.

This notion of community founded on an essence coincides with Vasconcelos's philosophical system in general and, as will be seen later, with the theory of the cosmic race in particular. It is meaningful, however, that in his thinking the plant and the jungle surreptitiously stage a contingent and proliferating multiplicity that, at the same time, is domesticated or subsumed into a stable totality. Similar to the implications of the plant's dependence on the elements, all semblance of unalterable identity is dismantled as soon as the fact that a single plant contains in itself a potentially infinite heterogeneity is taken into account. Vasconcelos states: "Dividing an animal, Fabre continues, is tantamount to destroying it. The individual is a creature, a living unit: the dog, the cat; the plant, however, lives collectively. Dividing vegetal life is the same as multiplying it; by sectioning a vineyard, the winemaker multiplies his vineyards; the rosebush is propagated through vine shoots" ("Dividir el animal, sigue comentando Fabre, es destruirlo. Lo individual está constituido por una criatura, una unidad viviente: el perro, el gato; la planta, al contrario, tiene vida colectiva. Dividir la vida vegetal es multiplicarla; seccionándolos, el viñador multiplica sus viñedos; el rosal se propaga por sarmientos").[101] Unlike animals, plants have no biological or ontological identity, not only because they depend radically on the external elements, but also because their body parts are not ultimately determined by an organizing center and could even function as autonomous individuals if necessary. Each plant hosts a multitude of others whose relations with each other are neither definitive nor necessary in biological or ontological terms. That is why, for example, Deleuze and Guattari derive their concept of rhizome by observing certain plants whose roots grow without a center or hierarchy and establish unexpected connections with each other.[102]

Vasconcelos unveils and simultaneously conceals the emergence of such a multiplicity of differences because his thinking primarily seeks to discover the principle that provides a sound foundation for community. If every community is ultimately a being-in-common of immeasurable (it cannot be divided, classified, or structured in a definitive way) and untotalizable (it cannot be unified or synthesized in a totality) nature, on the contrary, Vasconcelos strives to discover the "architectural plan" that supposedly arises organically from the community/jungle and contains its immanent multiplicity. Such a "plan" can be thought of as a transcendent entity that distributes the social roles and subsumes the common in a stable configuration. Thus, the jungle as a model of botanical ethics would imply that human communities must aspire to a harmonic and organic existence, without possible fissures or conflicts, which in Western political philosophy has taken formulations such as Rousseau's "the state of nature" or the communitarianisms of different ideological signs. By the 1930s, precisely when Vasconcelos was writing his theory on botanical ethics, Nazism was propounding the concept of a nation whose bright future depended on organic homogenization and purity of blood. It should not come as a surprise, thus, that Vasconcelos supported Nazi ideology and propagated anti-Semitic ideas through his journal *Timón*.[103]

As already mentioned, Vasconcelos—in keeping with the dominant tradition of Western metaphysics—conceives the seed as the generating power whose teleological fulfillment engenders plants and all beings. But one could say that in the seed—the symbol of the *arkhé*—there is always already an ontological multiplicity that deterritorializes identity. The Mexican intellectual even refers to the "infancy" of plants, a period characterized by its radical contingency: "Plants do not change places, and elements come to it: but their germ travel. What perfect beings are these that move in their infancy but that once they have overcome the mystery of (re)generation they settle forever and carry out their development with the serenity of a being in control of its environment?" ("La planta no cambia de lugar, y a ella acuden los elementos; pero sus gérmenes caminan. ¿Qué seres perfectos son estos que se mueven en la infancia, pero que, una vez que han vencido el misterio de la generación, se establecen para siempre y consuman su desarrollo en la paz de quien domina su ambiente?")[104] The fact that the "germs" or seeds "travel" in search of "the mystery of (re)generation" implies of course the possibility of an initial failure to blossom, as well as the possible accidents related to the place and circumstances of germination, growth, etc. In other words, the contingent "movement" of the seeds suggests that the ontological program supposedly inherent in themselves is neither definitive nor necessary. This is what, in relation to the concept of dissemination, Derrida calls the "heterogeneity and absolute exteriority of the seed."[105] There is, then, a multiplicity of "destinies" or "tasks" that gather within seeds, which is equivalent to saying that beings do not have a specific ontological vocation. Of course, Vasconcelos recognizes such ontological multiplicity but, at the same time, confines it exclusively to a provisional period, infancy, in order to neutralize it. He thus favors the image of plants that "have overcome the mystery of (re)generation" and "settle forever" in a definitive ground.

THE COSMIC RACE

Canonized as a foundational work in the Mexican and Latin American intellectual tradition, La raza cósmica has generally been analyzed as a "utopian essay" ("ensayo utópico") that postulates a historical teleology based on a notion of race both biological and spiritual.[106] Critics have highlighted that the idea of the future fusion of all races may not be taken literally as a prophecy, but nonetheless it has had certain ideological repercussions in different historical contexts: while, for example, in postrevolutionary Mexico, Vasconcelos's ideas served to legitimize the political and racial incorporation of Indigenous communities to the modern state, within the Chicano activist groups, his theses were construed as a multiculturalist proposal that exalted the cultural hybridization of Latin American societies.[107] In general, critics have underlined that the ideological project of *"mestizaje"* condones a biopolitical order that, despite its decolonial intentions, ends up reproducing the Western racial hierarchy and neutralizes corporeal differences and Latin America's social diversity.[108] Thus, it has been widely shown how La raza cósmica formulates a totalizing narrative that, departing from a Hindu or Neoplatonic monism, strives to homogenize and purify racial, political and ontological heterogeneity.[109]

Less familiar is the suggestion that Vasconcelos supports his utopia of racial miscegenation on the pea plant hybridization experiments conducted by Gregor Mendel.[110] As is well known, Mendel's principles of inheritance—as well as the work of August Weismann—provided the scientific grounds for "hard" eugenic policies in countries such as the United States and Germany, while the neo-Lamarckian theory of acquired characteristics inspired the kind of "soft" eugenics found in Mexico, France, and Brazil.[111] Mendel's discoveries were typically construed as the rationale for the crucial importance of untouched desirable genes and, therefore, the necessity of avoiding or prohibiting any breeding with "lower" genes/races. However, in *La raza cósmica* Vasconcelos presents an opposite interpretation of Mendel's experiments, arguing that they prove the fact that hybridization between antithetical types tends to create better individuals.[112] Furthermore, Vasconcelos brings Mendel's research to bear on his own spiritual interests, "combining the improvement of genes with the improvement of the soul," as Susan Antebi puts it.[113] As opposed to animal-inspired theories such as Darwinism, the fact that Mendel worked on plants turns out to be crucial for Vasconcelos's project of counteracting the idea of "survival of the fittest." This is why, as Ana María Alonso has contended, in *La raza cósmica* and elsewhere the Mexican intellectual often uses "organic imagery of plant growth as well as the trope of grafting as an intentional technique for stimulating hybrid vigor."[114] According to Vasconcelos, plant grafting provides a more appropriate image than animal breeding for envisaging how the Cosmic Race will integrate and uplift all the previous races.

La raza cósmica is without a doubt a text that is fully inserted in the intellectual coordinates that Vasconcelos establishes in his philosophical works: it is a teleology of ascending or spiritualizing nature that goes through moments of transformation with the ultimate goal of integrating differences into the One called spirit or Absolute. As in Vasconcelos's account of the hierarchy of lifeforms, *La raza cósmica* similarly presents an ambivalent conceptualization of each successive race or "social stage" ("estadio social"): each of them are to be considered, at the same time, as realities that have intrinsic value and as stumbling blocks that must disappear to attain higher forms of being.[115] Ultimately, though, the Cosmic Race will be the result of the "transcendental plan" that prescribes the homogenization of all individual parts to form a harmonic and elevated totality. The last race will closely follow the model of collectivity that Vasconcelos theorizes through the figure of the jungle: a community founded on a "destiny" or "trascendental mission" that provides an ontological and historical justification.[116] One could say that the Cosmic Race, "capable of true fraternity and truly universal vision" ("capaz de verdadera fraternidad y de visión realmente universal"), will adopt the eminently collaborative behavior of plants to establish a kingdom of harmony in which there will be no conflicts or social evils: "poverty, defective education, the scarcity of beautiful types, the misfortune that makes people ugly, all these calamities will disappear from the future social state" ("La pobreza, la educación defectuosa, la escasez de tipos bellos, la miseria que vuelve a la gente fea, todas estas calamidades desaparecerán del estado social futuro").[117] Just as in the jungle different plant species collaborate with a common goal over time, the various historical races have fulfilled their missions and contributed to the purpose of creating a synthetic totality that will transcend biological limitations in search of the highest stages of spiritual development.

However, the jungle not only serves as a model of behavior in *La raza cósmica*, but also as a natural environment that facilitates the splendor of the last race. Vasconcelos suggests historical, economic, environmental, and even alimentary grounds in order to support his conviction that the Cosmic Race will dwell in a metropolis called Universópolis located in the tropical jungle of South America. Well known is the fact that the historical narrative of *La raza cósmica* rests on the assumption—derived from (pseudo) scientific theories such as the esoteric myth of Atlantis—that humanity first emerged in the tropical jungle and will necessarily return to the same place at the end of its transcendental mission.[118] It is a cyclical conception of history according to which historical time has been shaped by civilization cycles led by a particular race: "There is no way back in history because it is all about transformation and novelty. No race comes back; each one sets out its mission, accomplishes it and disappears" ("En la Historia no hay retornos, porque toda ella es transformación y novedad. Ninguna raza vuelve; cada una plantea su misión, la cumple y se va").[119] The White race has already fulfilled its mission of "mechanizing the world" ("mecanizar el mundo") and, therefore, it has established the conditions for the definitive human regeneration in Latin American territory.[120]

By reversing dominant ideas held by social Darwinists such as Herbert Spencer, Vasconcelos argues that warm climate was the cradle of the most sophisticated civilizations and will be the perfect environment in which the ultimate race will flourish. In his *Ética*, he claims that historically the first great civilizations developed when they left hunting behind and started to sustain their way of life in agriculture: "The true civilization started in the Hindu or Mayan tropics, took grain from plants, without any need for ritual, the carnivore's skull breaker. And agriculture consolidates an ethical stage within the savage era rooted in hunting" ("la verdadera civilización se inició en el trópico hindú o maya, tomando de las plantas el grano, sin necesidad del rito, rompecráneos del carnívoro. Y la agricultura consolida una etapa ética dentro de la era salvaje asentada en la cacería").[121] The intimate relationship between human beings and plant crops produced the "true civilization," in opposition to the savage habits of human beings devoted to hunting. Agriculture engenders a form of sedentary and sophisticated civilization that allows the development of the spiritual and intellectual faculties most valued by Vasconcelos. Thus, in search of spiritual ascent the Cosmic Race will adopt the habits of tropical plants that are observed both in the jungle's "architectural plan" and in the civilizing influence of crops.

Vasconcelos even goes further and suggests that the definitive race will appear when humanity not only embraces botanical ethics, but also adopts a thoroughly vegetarian diet: "This whole contemporary civilization founded more or less English-style on animal husbandry and breeding, could regenerate itself if we switch a little closer to botany, seeking in it nourishment derived from fruits and grains" ("Toda esta civilización contemporánea, fundada, más o menos a la inglesa, en la engorda y cría de animales, podría regenerarse si nos pasásemos un poco a la botánica, buscando en ella la alimentación derivada de frutos y granos").[122] Future biological sciences, breaking free from the unproductive study of morphological aspects, will focus on unraveling "the mystery" of "the tropical fruits, which someday will revolutionize food hygiene" ("misterio [de] las frutas del trópico, que algún día revolucionarán la higiene alimenticia").[123] It is known that Vasconcelos's interest in vegetarianism arose from his readings on yogic philosophy, which even had a clear impact on the educational campaigns of postrevolutionary Mexico.[124] In *Estudios indostánicos* (1920), he had already argued that vegetarianism and warm weather are two constant factors in all spiritually sophisticated civilizations, while barbarian cultures tend to be carnivorous and live in cold climates (186). Historically, "men of greater mental energy have been, almost without exception, vegetarians, and that the most intelligent peoples are more or less vegetarian" ("los hombres de mayor energía mental han sido, casi sin excepción, vegetarianos, y que los pueblos más inteligentes son más o menos vegetarianos").[125] The definitive race will therefore have an exclusively vegetarian diet with the aim of promoting the development of the "the higher standards of sentiment and fantasy" ("normas superiores del sentimiento y de la fantasía").[126]

Therefore, according to Vasconcelos, the tropical regions of Latin America possess all the environmental characteristics necessary to accommodate the last race: "natural resources, arable and fertile lands, water and climate" ("recursos naturales, superficie cultivable y fértil, agua y clima").[127] It is important to note here that the theory of the Cosmic Race is ultimately based on a socio-ecological framework, that is, a proposal for managing the relationship between society and nature. From the outset, it is clear that *La raza cósmica* reproduces the notions substantiated both by the history of Western philosophy and by capitalism as a socio-ecological regime— that human consciousness enjoys an ontological and political privilege over nature understood as an independent sphere over which humans exercise unrestricted control.[128] The political and economic power of the Cosmic Race will arise precisely from the technological appropriation and efficient exploitation of the vast unexplored resources of the South American tropics, from arable land, to precious metals and timber, to the thermodynamic force of the Iguazu Falls.[129] "Nowhere else in the world" Vasconcelos claims, "are there promises of material wealth comparable to those of these regions" ("No hay en el mundo promesas de riqueza material comparables a las de estas regiones").[130] As Jason Moore has suggested, the cyclical renewal of capitalism requires the search for and conquest of new frontiers, that is, new spaces whose reserves of resources, energy, and unpaid labor allow the intensification of the capitalist accumulation process.[131] The utopia of Universópolis is clearly inserted in this capitalist logic: the South American tropics are the "last frontier" that will sprout humanity's definitive period of economic and political splendor.

So far I have argued that vegetal life occupies a preponderant place in Vasconcelos's utopia: the coming of the Cosmic Race involves a decisive relationship with plant life in instrumental, historical, and political/ethical terms. But there is one more crucial aspect in which the Cosmic Race will resemble vegetal life: the fact that plants serve as "cosmic mediators" that put into communication the ground and the air, the Earth and the sun.[132] According to Vasconcelos, "It is thus evident that we are receivers of universal energies in order to transmute its course and to accommodate them to the moral order first and then to the aesthetical" ("Se ve así patente que somos receptores de las energías universales con el objeto de transmutarles el curso y para acomodarlas al orden moral primero y después al estético").[133] In other words, humanity *collects* the primordial cosmic energy that created all that exists and then operates a *transmutation* in order to accomplish the last revulsion of the energy and reach the aesthetical/spiritual realm.[134] As Vasconcelos puts it, "unknowingly and knowingly, man performs a transcendental cosmic function, the function . . . as if it were to transport the entire creation to a different plane from the physical, perhaps an already immortal, divine one" ("el hombre desempeña sin saberlo y a sabiendas, una función cósmica trascendental, la función . . . como si dijésemos de transportar la creación entera a un plano diferente del físico y acaso ya inmortal, ya divino").[135] The Cosmic Race will reach what Vasconcelos calls the third period (spiritual or aesthetical), in which human nature "is satisfied and dissolves in emotion and is confused with the joy of the Universe: it becomes a passion for beauty" ("se satisface y se disuelve en la emoción y se confunde con la alegría del Universo: se hace pasión de belleza").[136] In a way, the last race will function as a "cosmic mediator" between the cosmic energy and the last revulsion, between the material world and the spiritual realm. Inspired by the dependence of plants on the cosmic environment (sunlight, air, soil), the last race will adopt the cosmic behavior of plants and accept the responsibility of overcoming the biological sphere.

Thus, the "transcendental cosmic function" of the last race is nothing less than fulfilling the ultimate plan of the universe. If the primordial energy prescribes an upward and transcendent direction, Vasconcelos suggests on several instances that one should understand this assertion not only in spiritualizing terms, but also in *literal* terms. When it reaches the final, spiritual stage, the energy "will extend the entire cosmos toward the open sky, where all dimensions break and grow, and a trembling rhythm saturated with infinity becomes wider" ("prolongará todo el cosmos hacia el cielo abierto, donde todas las dimensiones se rompen y crecen y se ensancha un ritmo que tiembla saturado de infinito").[137] Vasconcelos even suggests that biological life and the Earth are only fleeting settings in which universal life manifests itself. At some point, human societies and the planet as a whole will have to disappear or be destroyed in order to reach the highest stages of the Absolute: "society must become servant and recognize that it does not have its own view, but is a kind of transitory process, a way for men to conquer each one for themselves, but everyone in the end, the destruction of society, its disappearance as part of planetary life, precisely of the life that we want to annihilate in order to overcome it" ("la sociedad debe hacerse sierva y reconocer que no posee mira propia, sino que es una especie de proceso transitorio, una manera de que los hombres conquisten cada uno por sí, pero todos al fin, la destrucción de la sociedad, su desaparición como parte que es de la vida planetaria, precisamente de la vida que queremos aniquilar para poder superarla").[138] In consonance with other biocosmists' interest in space exploration, Vasconcelos states that the

Cosmic Race is bound to *overcome* earthly life and flee from Earth toward the infinity of the cosmos, since in the end the human spirit "comes from the earth and wants to get away from the earth" ("procede de la tierra y quiere alejarse de la tierra").[139]

This is why architecture in Universópolis will take the form of a spiral that penetrates the cosmos. The last race will build "spiral buildings, because the new aesthetics will try to conform to the endless curve of the spiral that represents free longing; the triumph of being in the conquest of infinity" ("construcciones en caracol, porque la nueva estética tratará de amoldarse a la curva sin fin de la espiral que representa el anhelo libre; el triunfo del ser en la conquista del infinito").[140] The Cosmic Race represents the highest form of universal life and will overcome, while also encompassing, all the previous forms of existence in the cosmos. The Cosmic Race showcases "a yearning for infinite integration and totality that therefore invokes the Universe" ("una ansia de infinita integración y de totalidad que por lo mismo invoca al Universo").[141] Thus, the last race aspires to bring together and surpass not only all existing races, but also everything on Earth and throughout the infinite cosmos created by the primordial energy. It will eventually comprise the whole universe, just like "the spiral that ascends encompassing each step more Universe, fleeing from the sphere, drilling upward as if to flee from space" ("la espiral que asciende abarcando a cada paso más Universo, huyendo de la esfera, perforando hacia arriba como para huir del espacio").[142] In a way, once the definitive race comes to fruition, the primordial energy will retrace its steps and reach back to its first environment: the cosmic space. As already seen, this will constitute the last stage of the creative force, when the primordial energy will finally continue its unrestrained movement for infinity.

In this sense, an examination of Vasconcelos's ideas on plants and botanical ethics casts an unusual image of the Cosmic Race and opens up an analytical perspective that has rarely been explored. Mostly focused on the issue of race and the deconstruction of *mestizaje*, a vast majority of critical studies disregard the fact that the racial order proposed in the text is based on a socio-ecological organization in which vegetal life has instrumental (natural resources, nourishing diet), historical (importance of agricultural civilizations) and political/ethical (ideal model of community, cosmic mediation) significance. Specifically, the critical perspectives focused on racial issues tend to overlook the properly cosmological aspects of Vasconcelos's utopia. The Cosmic Race will not only live in the jungle and eat plants exclusively, it will *be* like plants in political and ethical terms. More importantly, the

Cosmic Race will adopt the cosmic behavior of plants for fulfilling the ultimate purpose of the universe.

CONCLUSION

Vasconcelos's thought was primarily concerned with grasping the cosmic forces that created and unify all that exists, from outer space to the earthly beings to the unknown beings populating other planets. Since his initial publications, Vasconcelos strove to build a philosophical system soundly underpinned by the creative cosmic energy that, through its various revulsions, pursues its destiny of creating the human consciousness and the highest spiritual realm. Vasconcelos's engagement with life across the universe entailed a consideration of the ethical behavior of nonhuman agents, including inorganic matter and living organisms. He centered on beings—especially horses and plants—that decidedly display the upward direction of the energy toward the Absolute. Regarding vegetal life, in line with Kahlo's exploration of human-plant hybrids, Vasconcelos highlighted the cosmic function of interconnecting the biological, geological, and cosmological realms. His theorization of plant consciousness and collectivity as the basis of botanical ethics opens onto an innovative rereading of the Cosmic Race utopia, which shifts the focus from a merely racial perspective to one attuned to its cosmological and environmental implications. Viewed in this context, the theory of the last race is founded on a socioecological organization that has an intimate relationship with the destiny of the whole cosmos.

Vasconcelos's speculations on plants and botanical ethics are one of the most unexplored and intriguing areas in his entire body of work. It is also a set of suggestions that ultimately display the fundamental tension that I argue runs through Vasconcelos's thought and biocosmism at large: the tension between anthropocentric, foundationalist structures and a destabilizing, posthuman impulse. On the one hand, the sensitive and conscious behavior of plants gives rise to the notion of consciousness as a tight interweaving of organic and inorganic elements. This idea of plant consciousness unsettles the ontological, political, and environmental privilege of the human being promoted by Vasconcelos's own thought. On the other hand, the collective nature of the plant reveals a proliferating and immeasurable heterogeneity, which he himself seeks to subsume into a stable totality. But the multiplicity within the community and within "the seed" itself tends to disrupt any attempt at stabilization. In this way, departing from the notions of plant consciousness and collectivity, one can clearly unravel the political and philosophical shortcomings of Vasconcelos's thought and his theory of the Cosmic Race. Perhaps more importantly, his encounter with plants—his attempt at listening to "their silent language [that] barely touches our crude consciousness" ("lenguaje silencioso [que] roza apenas nuestra burda conciencia")—suddenly uncovers the possibility of postfoundationalist and non-anthropocentric thought.[143]

CHAPTER 4
Dr. Atl and Nahui Olin
Volcanism, Cosmological Forces,
and Space Exploration

The personal and creative partnership between Dr. Atl (Gerardo Murillo, 1875–1964) and Nahui Olin (Carmen Mondragón, 1893–1978) produced one of the most recognizable set of biocosmic ideas in postevolutionary Mexico. Dr. Atl and Nahui Olin met in 1921 and engaged in a deep romantic and intellectual relationship that lasted approximately two years. At the time, Dr. Atl had already spent a few years in Europe studying painting and philosophy and participating in early avant-garde groups in France.[1] Once back in Mexico, he was one of the firsts advocates of the Mexican Mural Movement and worked as a political organizer for the Constitutionalist faction during the Revolution. Over the course of the next decades, Dr. Atl engaged in a wide variety of intellectual projects as a visual artist, writer, philosopher, volcanologist, city planner, among others. By the time of his death, he had established himself as one of Mexico's most recognized visual artists of the twentieth century. He was buried in the Rotonda de las Personas Ilustres in Mexico City and his artwork was officially declared national heritage by the Mexican government.[2] For her part, in 1921 Nahui Olin was an artist who had also spent time in France where she became involved in the art scene. In the 1920s and 1930s she published several books of poems, created paintings and drawings, and worked as a model for artists such as Diego Rivera and Edward Weston. After gaining recognition as a multifaceted artist, Nahui Olin disappeared from public life in the 1940s and led an inconspicuous existence until her death. Posthumously, since the 1990s she has become the focal point of large solo exhibitions, biographies, critical studies, and even novels and films, emerging as a well-known figure of postrevolutionary culture.[3]

The partnership between Dr. Atl and Nahui Olin was marked by an intense intellectual dialogue nurtured by a shared scientific outlook and a similar utopian drive. During the 1920s and 1930s, even after they had terminated their personal relationship, their intellectual projects took similar paths by addressing the same cluster of issues concerning the natural world and the universe. Both of them were captivated by new theories of cosmology, electromagnetism, and relativity that expanded the standard understanding of the universe. For example, whereas Dr. Atl was fascinated by Shapley's estimations of the scale of the universe, Nahui Olin was interested in engaging with Maxwell's theory of the electromagnetic field and Einstein's theory of special relativity. At the same time, Dr. Atl and Nahui Olin were passionate about reading speculative fictions such as Flammarion's novels concerning space travel, new scientific discoveries, and so on.[4] In their own work, Dr. Atl and Nahui Olin built on novel scientific theories with the aim of speculating about the ultimate constitution and limits of the universe, as well as theorizing humanity's role in conquering and transforming the cosmos. Biocosmism's fascination with space exploration and colonization found its maximum expression in their work. As I will argue in this chapter, both artists produced utopian theorizations that highlighted the indissoluble links between human beings, the Earth, and the cosmos.

I will first delve into Dr. Atl's understanding of the vitality of the Earth and its interrelation with humanity and the cosmos through a reading of his work related to telluric phenomena. I will mainly examine his early literary account of volcanoes *Las sinfonías del Popocatépetl* (1921) as well as his later scientific work *Cómo nace y crece un volcán: El Paricutín* (1950), along with a selection of his landscape works. The second part will focus on Nahui Olin's *Óptica cerebral* (1922) and *Energía cósmica* (1937) to shed light on the notion of an expanded, direct interaction between cosmological forces and human beings. I will focus on Nahui Olin's suggestions for radically transforming the universe by destroying its uniform and harmonious stability. In the third section, I will establish a dialogue between Nahui Olin's ideas and Dr. Atl's own speculations about humanity's ascending thrust through an analysis of his space travel novel *Un hombre más allá del universo* (1935). This section delves into the fundamental contradiction that lies at the core of Dr. Atl's and biocosmism as a whole: a tension between the desire to master the universe through space travel, and the permanent drive to discover the limits of the cosmos and human knowledge. Lastly, I will explore how this same contradiction manifested itself—and seemed to be resolved—in the utopian project of

Olinka: a city inhabited by scientists and intellectuals that would consummate the ideal of reaching the stars.

THE VITALITY OF THE EARTH

Dr. Atl's engagement with diverse geological phenomena is a well-known aspect of his eventful life and wide-ranging work. Throughout his life he was involved in various ways of exploring Earth's features as a devoted mountaineer, landscape artist, volcanologist, and leader of projects concerning the exploitation of mineral resources. Dr. Atl prided himself on spending long periods of time in the wilderness, camping at the foot of mountains and volcanoes across central Mexico, where he would draw and paint landscapes in a highly personal style. He crafted a mythology about himself centered on his involvement with the natural world: the walking geologist/artist who is drawn to living in the wilderness, the artist/philosopher who decided to adopt Atl (Nahuatl for "water") as his artistic name in the middle of a storm. As Peter Krieger contends, art scholars have traditionally relied heavily on this mythology to interpret Dr. Atl's work, reproducing his words and ideas in an acritical manner, which has resulted in the repetition of commonplace remarks and a concealment of his most problematic facets, including his Nazi sympathy.[5] Additionally, while his literary works have been for the most part disregarded, his visual landscapes and his overall persona have been embraced as a central component of *mexicanidad*: the national identity purportedly expressed and revived by postrevolutionary artists.[6] As I will show, Dr. Atl's landscapes have been mostly analyzed by inscribing them in the nationalist visual tradition epitomized by José María Velasco.

In contrast with these recurrent interpretative lines, Peter Krieger has proposed to establish a dialogue between Dr. Atl's works and the so-called geological turn in the humanities and social sciences: the recent outburst of interest in the interpenetration between geological processes and the artistic, cultural, political, and socioeconomic features of societies.[7] Krieger advances the idea that Dr. Atl practiced a sort of "geo-graphy" or "Earth-writing" that engaged in various ways "the fascinating metamorphoses of the planet's lithosphere and atmosphere in the diminutive territory circumscribed by the Mexican highlands," which bear the mark of ancient, ongoing geophysical forces.[8] According to Krieger, Dr. Atl strove to accomplish a "transformation of landscape's telluric and atmospheric energy" into images that could affect spectator's bodies and imaginations through a potent visual experience.[9] In this vein, I will argue that Dr. Atl endeavored to produce a bodily engagement with what Elizabeth Grosz calls "geopower" or the forces of the Earth, which "while 'unliving' as chemical elements and forces, can be understood as having a kind of life of its own when it is understood as a system of order and organization that is continually changing, never fully stable, dynamic."[10] Additionally, I will suggest that behind Dr. Atl's obsession with geological phenomena—particularly volcanoes—lies a *cosmic* rationale that often goes unexplored in the critical assessments of his work, including in Krieger's suggestive reading. This cosmic understanding ties together two areas of Dr. Atl's initiatives that are regularly considered unrelated and even diametrically opposed: his interest in volcanoes and his passion for cosmological speculations.

Dr. Atl's little-known participation in gold mining can provide us with initial insight regarding what was at stake in his overall engagement with earthly forces.[11] In the 1930s, he became involved in prospecting and exploiting oil and gold in Mexican territory, with varying results, while also producing historical works that documented and publicized these initiatives.[12] Dr. Atl imagined Mexico as a country full of untapped mineral resources that he would finally prospect and exploit for the immense benefit of the people. Even though Mexico had never excelled at gold production, Dr. Atl was convinced his native country could become one of the world's largest producers of this precious metal. He drafted a comprehensive plan for exploiting it in several locations with the hope of gaining the financial support of the Mexican government, which eventually implemented a limited version of his ambitious plan.

Dr. Atl believed that Mexico's gold production could contribute to recovering from the worldwide economic depression of the 1930s. But he was also convinced that humanity's overlong fascination with gold was not merely based on this utilitarian, economic motivation. In his work *¡Oro! Más oro: El mundo lo necesita, Méjico puede dárselo* (1936), Dr. Atl speculated that gold has always been deemed the most valuable and beautiful metal because it exercises some sort of unknown material attraction—similar to the attracting force of gravity or electromagnetism—that human beings and all kinds of bodies cannot resist. As he argued: "Gold is an immense magnet whose potentiality in gauss, distributed throughout the globe, acts on bodies—and on man—creating a magnetic phenomenon from which the human species has not been able to escape" ("El oro es un inmenso imán cuya potencialidad en gauss, distribuida en todo el globo, actúa sobre los cuerpos—y sobre el hombre—creando un fenómeno magnético al cual la especie humana no ha podido sustraerse").[13] In this sense, far from only having aesthetic or utilitarian properties, gold's composition must contain a kind of material vibrancy that unknowingly attracts and transforms bodies.

This speculative theory of gold's material attraction constitutes an example of Dr. Atl's comprehensive engagement with geological processes and features. Inspired by Maxwell's theory of the electromagnetic field, Dr. Atl built on the premise that the Earth relies on a vital vibrancy that has a material impact on all beings. This general framework is clearly at work in Dr. Atl's engagement with volcanism, which is one of the most prominent aspects of his work. Understandably, he had a strong and meaningful connection to the Popocatéptl and the Iztaccíhuatl, distinguishing geological features of the Valley of Mexico that had appeared extensively in the landscape works of nineteenth-century painters such as José María Velasco and Eugenio Landesio. But Dr. Atl also had the opportunity of admiring and studying firsthand the birth of the new volcano Paricutín in the state of Michoacán in 1943. He created visual, literary, and scientific accounts of these volcanoes coded as powerful and energetic signs of Earth's vitality that directly affect human beings. As I will show, Dr. Atl was fascinated by the "conscious" and methodic behavior of volcanoes that revealed the existence of *cosmic* forces and tendencies.

In *Las sinfonías del Popocatépetl*, Dr. Atl provides a literary account of his interaction with "la montaña viviente" ("the living mountain") as he calls the famous volcano that gives its name to the work, and of how this intense interaction shaped—not only ideologically, but also *materially*—his way of understanding and dwelling on the Earth.[14] In contrast with allegorical readings of this work that view the volcano as the symbol of the revolutionary event and the interruption of historical time, I would like to analyze Dr. Atl's material engagement with geopower.[15] The Popocatépetl reveals itself throughout the work as a lively manifestation of telluric energy, which has undergone a constant process of transmutation since the formation of the Earth. For starters, the Valley of Mexico bears the geological marks of the volcanic phenomena that shaped the region: on his way to the Popocatépetl, Dr. Atl witnessed "extinct craters that one time vomited the ashes that littered the great lakes of Anáhuac and formed the subsoil of this immense valley" ("cráteres apagados que en un tiempo vomitaron las cenizas que azolvaron los grandes lagos de Anáhuac y que forman el subsuelo de este valle inmenso").[16] Going even further back in time, Dr. Atl asserts that by standing on top of the mountain and staring at the sky one can get an intuitive understanding and a broad image of the cosmic events that originated the Earth and everything in it: "Walking along the abrupt ledge of an icy mountain—the radiations of thought projected like the light of a beacon over the depths of Space—I felt the boundless importance of the insignificant and prodigious Accident that tore the clouds, and discovered the Worlds, and transformed Life into an endless spiral" ("Caminando por el áspero perfil de una montaña helada—las radiaciones del pensamiento proyectadas como la luz de un faro sobre las profundidades del Espacio—yo sentí la importancia sin límites del Accidente insignificante y prodigioso que rasgó las nubes, y descubrió los Mundos, y transformó la Vida en una espiral sin fin").[17] In this sense, the telluric energy that manifests itself in volcanoes constitutes in turn the manifestation and prolongation of the cosmic energy that created the planets, including living and nonliving beings on Earth.

Moreover, just as Dr. Atl emphasizes how the Popocatépetl is the product of long-standing and constantly diversifying cosmic energies, he also shows how this volcano—being an "active" volcano in geological terms—never ceases to transform in terms of shape or activity and will continue to do so in the future. Perhaps this permanent process of change is more clearly evident in the volcano's wide range of activity from long dormancy periods to intervals of violent eruption of gases and/or magma. Dr. Atl operates a personification of the Popocatépetl with the aim of drawing attention to its changing degree of activity throughout the times: "Mausoleum of terrestrial energy, during centuries you slept on the mountains, fertilizing immense plains in the rest of your silence. But one day, the blood of the Planet, which boiled in your hidden arteries, flowed to your frozen heart, and the ancient fury of your incandescent eloquence came out of your extinguished mouth, announcing to the terrified people your prodigious resurrection" ("Máusoleo de la energía terrestre, siglos dormiste sobre los montes, fertilizando en el reposo de tu silencio inmensas llanuras. Pero un día, la sangre del Planeta, que hervía en tus ocultas arterias, fluyó a tu helado corazón, y salió por tu apagada boca la antigua furia de tu elocuencia incandescente anunciando a la aterrada gente tu prodigiosa resurrección").[18] In this quote, the Popocatépetl has become a dormant terrestrial body—akin to a mausoleum housing latent telluric energy—that at a certain moment in time "resurrects" by transporting the Earth's blood through its veins and violently spewing it out. These volcanic eruptions are a distinctive and enthralling expression of Earth's vitality: the preferred way in which the planet releases "the burning consciousness of Earth" ("la conciencia ardiente de la tierra"), as Dr. Atl puts it.[19]

But even when the Popocatépetl is dormant and apparently inert, Dr. Atl recognizes in this volcano a great deal of lively behavior that resonates with Herrera's and Vasconcelos's investigation on the vital vibrancy of inorganic matter. For example, Dr. Atl highlights the fact that the Popocatépetl—far from being an immobile and unvarying body—is constantly changing due to its interaction with elements such as wind and rain. Most of these changes happen at a pace that makes it difficult for humans to perceive them with the naked eye. The minuscule and incessant erosion done by the wind to the mountain is one of the instances explored in *Las sinfonías del Popocatépetl*. As Dr. Atl points out:

> And one day the wind detached a microscopic particle from the new rocks, and another day, another, and another day. . . . The Mountain continued impassive and through the ages it still appeared unscathed. But time continued to pass and the wind as well, tireless, tenacious, fierce. . . . It smashed its stone spines, hammered its cliffs, turned them to dust, uprooted its forests, and with terrifying energy tore apart the crater.

> *Y un día el viento desprendió de las rocas nuevas una microscópica partícula, y otro día otra y otro día otra. . . . La Montaña continuaba impasible y a través de las edades aparecía todavía incólume. Pero el tiempo continuó pasando y el viento también, incansable, tenáz [sic], feróz [sic]. . . . Destrozó sus espinazos de piedra, martilleó sus acantilados, los convirtió en polvo, desenraizó sus bosques, y con aterradora energía despedazó al cráter.*[20]

Dr. Atl observes how the mountain's disaggregated particles are carried over by the rain into river streams that on their way to the ocean end up forming sandy beaches: "There it goes the Mountain" ("Allá va la Montaña").[21] The mountain, which was once thought of as a monumental and stable body, it is now considered a set of minuscule and erratic particles in constant becoming.

In addition to the work done by external elements, volcanoes in their own right have the potential to alter their shape and form significantly, due to the Earth's internal forces. The birth of a new volcano gave Dr. Atl a privileged opportunity to witness the Earth's capacities for self-organizing by way of a surprising "constructive consciousness" ("conciencia constructiva").[22] When the news came out in 1943 that a volcano was abruptly appearing in the state of Michoacán, Dr. Atl set up a camp near the location with the aim of bearing witness and recording in written and visual forms the vicissitudes around this telluric phenomenon. He ended up living intermittently on site between 1943 and 1948. The result of this experience was *Cómo nace y crece un volcán: El Paricutín*, a hybrid work that combines personal observations, memoir, interviews with Indigenous witnesses, drawings, images, and scientific formulations.[23] In this work, Dr. Atl registered the most perceptible realities of Paricutín's "active, flashing, powerful life" ("vida activa, fulgurante, potente")—such as volcanic eruptions of magma and ashes that engulfed the nearby town of Paricutín—but he also recorded processes that were more understated and open to interpretation and, as such, helped him reach broader conclusions about the Earth and the cosmos.[24]

In this sense, one of the facts that struck him the most was recognizing that the Paricutín followed a clearly defined pattern in its process of growth and formation: every time the cone of the volcano reached a perfect form, Earth's internal pressure and temperature would destroy the shape of the cone by opening up cracks and holes in order to release magma. After this destructive phase, the volcano would work methodically to fill these holes up and achieve a seamless cone-shaped body once again. According to Dr. Atl, this whole process cannot be simply deemed a random and unintentional procedure. On the contrary, this pattern is undoubtedly a deliberate plan that displays prominently the volcano's conscience, its agency and willingness to self-organize: "This work was carried out with rigorous method, obeying a clearly established *plan*" (este trabajo fue llevado a cabo con riguroso método, obedeciendo *a un plan* claramente establecido").[25] This is why Dr. Atl speaks of the volcano's *intentions* and its *abilities* to perform this methodical strategy: "Tonight's maneuver leaves no doubt as to the *intentions* of (the volcano) to rebuild the cone, nor about the growing skill that it displays" ("esta maniobra de hoy en la noche no deja lugar a duda sobre las *intenciones* que tiene [el volcán] de reconstruir el cono, ni sobre la habillidad creciente que despliega").[26] In the end, the systematically destructive and constructive work done by the Paricutín surpasses easily all building techniques employed by humans: "The column renewed its reconstruction system, in such an obvious way, with such a perfect technique and such great speed, that our human methods are incapable of achieving, neither in precision nor much less in magnitude" ("La columna renovó su sistema reconstructor, de una manera tan evidente, con una técnica tan perfecta y una tan grande rapidez, que nuestros métodos humanos son incapaces de alcanzar, ni en precisión ni mucho menos en magnitud").[27]

However, even if the volcano seems to be performing a deliberate plan, Dr. Atl is clear in arguing that such plan should not be considered a preestablished and teleological program that has been prescribed by a deity or by nature's internal logic. In the scientific part of the work, Dr. Atl wonders: "Does this mechanism obey a pre-established order? What order? The one established by a divinity regulating all of the events of the Universe? Or is it one of many movements of an autonomous and unconscious mechanical system?" ("¿Obedece este mecanismo a un orden preestablecido? ¿A cuál orden? ¿Al establecido por un divinidad reguladora de todos los eventos del Universo? ¿O es uno de tantos movimientos de un sistema mecánico autónomo e inconsciente?") and later on he continues wondering: "Is nature a master of order, of method, of perfection? Is there in all the phenomena of the Universe a consciousness similar to that of Paricutín?" ("¿Acaso la naturaleza es maestra de orden, de método, de perfección? ¿Hay en todos los fenómenos del Universo una conciencia semejante a ésta del Paricutín?").[28] Dr. Atl comes to the following conclusion:

> Given the facts, it is logical to link the destructive activity to the constructive mechanics of the volcano, establishing a ridiculously teleological theory of cause and effect. Nature has no purposes, nor does it obey the principles of any order. Nature does not work or allow to act according to such or such laws, or beyond those laws. Laws have been invented by humans and to them we subject the phenomena that without our presence on earth would be meaningless.

> *Ante los hechos es lógico relacionar la actividad destructiva con la mecánica constructiva del volcán, estableciendo una teoría de causa y efecto ridículamente teleológica. La naturaleza no tiene finalidades, ni obedece a principios de ningún orden. La naturaleza no obra ni deja obrar conforme a tales o cuales leyes, o fuera de esas leyes. Las leyes las hemos inventado los humanos y a ellos sujetamos los fenómenos que sin nuestra presencia sobre la tierra carecerían de significado.*[29]

In other words, according to Dr. Atl, Earth's inorganic vitality displays particular tendencies and potentials that lead it to behave *contingently* in certain ways: this is what he refers to as the volcano's conscience and intentions. Ultimately, these patterns of behavior are ways in which nature tends to *spontaneously* conduct itself and self-organize in the face of particular conditions, without recurring to divine or natural organizing first principles. It is only because of the human disposition to produce fixed laws out of nature's constant mutations that Paricutín's growth process could appear as the result of preestablished and teleological plans.[30]

It is important to emphasize that, for Dr. Atl, the tendencies of Earth's behavior are properly *cosmic* tendencies in the sense that they present themselves throughout the universe. As already suggested, the origin and formation of volcanoes can only be accounted for by relying on "an explanation of cosmic nature" ("una explicación de carácter cósmico") that builds on "Universal phenomena that are ever increasingly more proved. I am referring especially to extraterrestrial (phenomena), identical to those that have occurred and are occurring on Earth, and whose fundamental cause is a primitive nuclear fire" ("fenómenos universales cada día más comprobados. Me refiero especialmente a los [fenómenos] extraterrestres, idénticos a los que se han producido y se producen en la tierra, y que tienen como causa fundamental un fuego nuclear primitivo").[31] In this vein, witnessing the birth of a volcano represents for Dr. Atl the rare opportunity to glimpse into the formation process that the Earth itself must have followed in the beginning of its existence and that it is still following in its continued transformation. During his personal observations, after a particularly brutal eruption of magma and ashes that created a somber and otherworldly scene, Dr. Atl suggests that he might be witnessing the abrupt creation of a new world:

> It seems to me that I have fallen into another world. And, indeed, this is another world, a world that is being created, and since within it there is no scale of comparison and the transformations take place with fantastic speed, we end up acquiring the conviction that the earth is forming. Is it not like this? Are we not witnessing the manifestation of a complete but reduced telluric phenomenon?

> *me parece que he caído en otro mundo. Y, en efecto, este es otro mundo, un mundo que se crea, y como dentro de él no hay escala de comparación y las transformaciones se suceden con fantástica rapidez, acabamos por adquirir la convicción de que la tierra se está formando ¿Y realmente no es así? ¿No asistimos a la manifestación de un fenómeno telúrico completo aunque reducido?*[32]

Dr. Atl seems to suggest that, in spite of its restricted geographical conditions, the birth of the Paricutín stages the same cosmic propensities that resulted in the geological features of the Earth and other planets across the cosmos.

Furthermore, Dr. Atl's engagement with volcanoes is regularly coded as a cosmic experience that opens up a material rapport between the human body and the universe. Dr. Atl illustrates his immediate interaction with cosmological and atmospheric forces in several passages contained in *Las sinfonías del Popocatépetl*. On one occasion, he describes how large and menacing clouds engulfed the mountains during a thunderstorm. Since Dr. Atl was located in such a high position and the clouds were extremely low, he dared to introduce his arm in a cloud, causing hail to fall down all over his body. He remained submerged in the cloud for a long time as he observed incandescent red and green ball-shaped objects plummeting from the sky. The atmosphere became full of electrical vibrations, indescribable colors, and a generalized smell of ozone. All of a sudden, Dr. Atl felt compelled to embrace even more thoroughly the stream of energies coming down the sky: "A strange force made me stand on the rock in an unconscious movement of desire. Those strange electrical worlds exploded over my body with fiery and silent speed" ("Una extraña fuerza me hizo ponerme en pie sobre la roca en inconsciente movimiento de deseo. Sobre mi cuerpo estallaron con fúlgida y silenciosa rapidez aquellos extraños mundos eléctricos").[33] Similarly, on a separate occasion Dr. Atl reports being knocked down by a sudden gust of wind while he simultaneously heard something falling down by his side. Immediately after, as he puts it, "Something terrible and cold, like a snake, coiled itself around my neck" ("algo terrible y frío, como una serpiente, se enroscó en mi cuello").[34] The dramatic way in which Dr. Atl relates these almost supernatural experiences of corporal exposure are meant to highlight how he deals with—and seems to live off of—formidable and overpowering cosmological forces in an unmediated way.

In the same vein, Dr. Atl recounted in what ways reaching the top of the mountain transformed his understanding of the universe and reinvigorated his bodily processes: "From the top of the Volcano I saw the World as a wonderful spectacle and I loved it without reluctance, deeply, intensely . . . and from all things emanated a new force whose influence I had never felt before—a palpitation whose rhythm was born from each molecule of matter and vibrated on my nerves, with renewing energy" ("Desde la cima del Volcán yo vi el Mundo como un espectáculo maravilloso y lo amé sin reticiencias, profundamente, intensamente . . . y de todas las cosas emanó una fuerza nueva cuyo influjo yo no había sentido jamás— una palpitación cuyo ritmo nacía de cada molécula de la materia y vibraba sobre mis nervios, con renovadora energía").[35] This quote suggests that living in the wilderness establishes a deep-level, even molecular, closeness between humans and nature. According to Dr. Atl, such interrelation is not mediated by perception or thought: it constitutes a material experience between bodies. Ultimately, reaching the top of the volcano puts one in the position of an immediate and direct relation with the cosmological forces that come from every corner of the universe: "The entire Universe pours on the Volcano the imponderable flow of its stars—it rains light—it rains light from the Cosmos on the World" ("El Universo entero derrama sobre el Volcán el imponderable fluido de sus astros—llueve luz—llueve luz del Cosmos sobre el Mundo").[36]

In a similar way, Dr Atl's landscape works also suggest a direct interaction between geological phenomena and cosmic energies. After an early involvement with avant-garde aesthetics, Dr Atl's visual art consolidated

FIGURE 4.1. Dr. Atl's *Volcán en la noche estrellada*. Photo courtesy of Laura Cohen. © Reproducción autorizada por el Instituto Nacional de Bellas Artes y Literatura, 2024

through landscape artworks "employing some avant-garde strategies with a nineteenth-century spirit" ("que utilizan algunos recursos de la vanguardia con un ánimo decimonónico").[37] Particularly, Dr. Atl draws from an established nineteenth-century tradition of landscape painting in Mexico by incorporating elements such as the preferred object of representation—the Valley of Mexico—and certain visual orderings like wide, dramatic views including mountains, volcanoes, lush vegetation, cloudy skies, etc. Because of this, Dr. Atl's landscapes are often considered instances of a nationalist visual tendency that strives to represent Mexican identity.[38] However, while nineteenth-century artists such as José María Velasco created landscapes infused with nationalist meanings, Dr. Atl emphasized instead pristine natural landscapes that seem to belong to a geological epoch predating the existence of human beings.[39] This is achieved, first, by removing all sociocultural

FIGURE 4.2. Dr. Atl's *Cráter y la Vía Láctea*. Photo courtesy of Laura Cohen. © Reproducción autorizada por el Instituto Nacional de Bellas Artes y Literatura, 2024

references and narrative scenes that implicitly connect natural features with particular Mexican cultural or historical issues—a strategy used masterfully by Velasco and others.[40] Secondly, Dr. Atl put forward an artistic approach that suggested and produced visual and compositional affinities between the Earth and the cosmos. I will describe the main aspects of his approach by analyzing *El volcán en la noche estrellada* (1950) and refer to a few other examples of his visual production.

Dr. Atl created dozens of works portraying Paricutín's wide variety of phases and behaviors. While some of the artwork presented otherworldly images of hardened lava and ashes by using a low vantage point that offered a view of the surroundings, other works highlighted the violence of ongoing volcanic eruptions by representing massive lava fountains and flows with bright colors. In contrast, *El volcán en la noche estrellada* portrays the erupting Paricutín under the starry night using a high vantage point that does not call attention to the violence or the damage caused by the eruption. Instead, the high vantage point suggests the fairly small size of the volcano in relation to the high, surrounding starry sky. There is, however, a clear visual connection between the volcano and the cosmos: the column of smoke and fire emerging from the crater rises so high that it seems to reach the sky. Significantly, the orange fire spots emanating from the column blend in seamlessly with the yellow and orange stars in the sky. This

FIGURE 4.3. Dr. Atl's *Boca de volcán*. Photo courtesy of Laura Cohen. © Reproducción autorizada por el Instituto Nacional de Bellas Artes y Literatura, 2024

visual equation makes it difficult to distinguish geological from cosmic phenomena, emphasizing the interpenetration between the Earth and the cosmos and expressing "the painter's immeasurable and sublime desire to unite the earth with sidereal space" ("el deseo incommensurable y sublime del pintor por unir la tierra con el espacio sideral"), as Olga Sáenz argues.[41] In later works such as *Cráter y la Vía Láctea* (1960) and *Boca de volcán* (1958), Dr. Atl emphasized the aforementioned interpenetration even more noticeably by making volcanoes almost another part of the sky, as if they were stars themselves.

Furthermore, *El volcán en la noche estrellada* displays other important aspects of Dr. Atl's approach to landscape painting. First, it incorporates the curvilinear perspective theorized by Mexican art professor Luis G. Serrano in a 1934 book that included a prologue by Dr. Atl himself. This perspective was conceived as counterpart to the traditional perspective in landscape works, which is structurally comprised of straight lines that organize the objects in terms of proximity or distance. In opposition, the curvilinear perspective employs curved lines that create a "spherical representation of nature" ("representación esférica de la naturaleza") allegedly more akin to "the circular impression of our senses" ("la impresión circular de nuestros sentidos").[42] According to Dr. Atl, this perspective "corresponds to a more universal sense of nature and allows a broader spatial expression" ("corresponde a un sentido más universal de la naturaleza y permite una más amplia expresión espacial").[43] Dr Atl appreciates how Serrano's technique allows him to produce an encompassing, monumental, and universal/cosmic depiction of nature.[44] In effect, the fact that the skyline is a curve instead of a straight line "traces the spherical shape of the globe" and underlines the idea that geological features are ultimately universal phenomena.[45] Moreover, in this painting Dr. Atl incorporates an additional circular shape that appears above the curvilinear skyline, perhaps representing the energy emanating from a spherical space body. Thus, the painting suggests that the Earth is just one among many cosmic bodies that are bound by the same forces and energies. Above all, the Paricutín and all geological processes are the result of universal tendencies and cosmological forces that affect them in a variety of ways.

In addition to the curvilinear perspective, *El volcán en la noche estrellada* incorporates petroleum-based paints fabricated by Dr. Atl himself, which he called "Atl colors." The invention of these colors made of "ceras, resinas y petróleo" ("waxes, resins and petroleum") demonstrates Dr. Atl's commitment to experimentation, not only in terms of perspective but also in the realm of the material components of painting.[46] He endeavored to manufacture new paints that would be unalterable, durable, and would also allow him to superpose layers of paint to create new material and imaginative effects.[47] The "Atl colors" achieved this purpose: "The general result is a great luminosity and a great wealth of matter" ("El resultado general es de una grande luminosidad y de una grande riqueza de materia").[48] In line with Jussi Parikka's approach to the geological underpinnings of media, one could argue that Dr. Atl produced new material engagements with geological matter with the hope of transforming his artistic intervention.[49] As Peter Krieger puts it, the expressive vibrancy and density of the "Atl colors" "augment the sheer physical power of the image on the observers' retina" and thus constitute a suitable medium for displaying the Earth's vitality in full force.[50] Dr. Atl endeavored to directly transpose the "pulsing sensation of what has been seen" ("sensación palpitante de lo que se vio") in landscapes to canvas, so as to create artwork that would "vibrate" in similar ways as nature.[51] As mentioned earlier, Dr. Atl believed that certain matter, such as gold, could exert a physical attraction akin to electromagnetic forces. In this sense, beyond simply being an artistic medium, the Atl colors could allegedly exercise a material attraction that would affect the body of the spectator. After all, these color paints are the product of a mixture of organic compounds that come from naturally occurring phenomena on Earth. In a way, the nature of the artistic medium itself already comprises a manifestation of the Earth's vitality and transformational processes, including the "stratigraphy" of superposed layers of matter.⁵²

Dr Atl's visual, literary, and scientific engagement with volcanoes encapsulated his approach to the lively and energetic activity of geological processes. He marveled at the Earth's ability to spontaneously self-organize with a high level of precision, which he attributed to cosmic tendencies at work throughout the universe. According to Dr. Atl, interacting with volcanism opened up the possibility of a cosmic experience marked by the interpenetration between the Earth and the universe, and between human beings and cosmological forces. Dr. Atl was convinced that human beings could maintain an unmediated, material interplay with the vital vibrancy that traverses everything in the cosmos. As I will argue next, some of these insights were further developed in Nahui Olin's work.

THE COSMIC VITALITY OF NAHUI OLIN

Much like Dr. Atl, Naui Olin has acquired a mythical status as a figure that embodied in many ways the cultural renovation of postrevolutionary Mexico.⁵³ Allusions to her precocious intelligence, otherworldly beauty, mysterious eyes, gender non-conforming behavior, and sexual openness have become commonplace since her rediscovery in the 1990s. Most of the scholarly readings of her work have centered on her personal and artistic challenges to heteropatriarchal norms in postrevolutionary Mexico.⁵⁴ Interestingly, according to Elissa Rashkin and Viviane Mahieux, Nahui Olin's gender and sexual dissidence was often linked to a *cosmic* aspiration: she explored "her own sexual being not so much in relation to a partner but as a way of connecting with the universe itself, blurring any border between body and mind" ("su ser sexual no tanto en relación con una pareja sino como una vía de conexión con el universo mismo, desdibujando cualquier frontera entre cuerpo y mente").⁵⁵ As Mariano Meza Marroquín puts it, she assumed "a kind of dynamism between man and the cosmos" ("especie de dinamismo entre el hombre y el cosmos") in which humanity affects, and is affected by, cosmological forces and energies in a direct way.⁵⁶ By reading Nahui Olin's works in the context of Mexican biocosmic thought, I aim to contribute to a deeper understanding of her cosmological discourse and its implications. Ultimately, I will argue that Nahui Olin endeavored to imagine speculative ways not only of understanding the universe, but also for radically transforming it.

Nahui Olin published five works during her lifetime: *Óptica cerebral, poemas dinámicos* (1922), *Câlinement je suis dedans* (1923), *À dix ans sur mon pupitre* (1924), *Nahui Olin* (1927), and *Energía cósmica* (1937).[57] The two volumes written in French are intimate, autobiographical poetry collections that stand out for their direct and unconventional exploration of corporality and sexuality.[58] For its part, *Nahui Olin* is a brief pamphlet in which the author gives an explanation for her pseudonym, suggested by Dr. Atl, which refers to the Aztec cosmogonic concept of the fifth and current sun, or age. In line with her cosmogonic pseudonym, the rest of Nahui Olin's work is characterized by a poetic and philosophical treatment of scientific themes related to physics and cosmology. Rather than strictly scientific treatises, *Óptica cerebral* and *Energía cósmica* are focused on speculatively imagining the philosophical ramifications of scientific concepts: "The writings of Nahui Olin are poetic approaches to scientific knowledge that try to place themselves between the real and the possible" ("Los escritos de Nahui Olin son acercamientos poéticos al conocimiento científico que tratan de situarse entre lo real y lo posible"), as Rebeca Julieta Barquera Guzmán and Mariana Rubio de los Santos suggest.[59]

Nahui Olin developed a philosophical account of the structure and ultimate purpose of the universe as a whole, including a reflection on the fundamental role reserved for the human being. In her work, the universe is conceived as the infinite and harmonious totality of existing and potential processes and objects.[60] This totality has no beginning and no end and, even though it is constantly transforming, its overall structure remains practically the same throughout the ages. By virtue of being interconnected, the entirety of cosmic objects and processes transform and mutually affect each other, building a harmonious equilibrium at each point in time. This dynamic yet unchanging quality is what Nahui Olin describes as "the vitality of the cosmos." In the end, all cosmic phenomena—physical and spiritual entities alike—is entirely dependent upon material processes such as forces and movements. In turn, this set of forces—some of which are known such as electricity and radioactivity, but others remain unknown— are ultimately manifestations of a unique current of cosmic energy, which is the primordial cause of everything that exists in the universe. Paradoxically, as I will show in this section, the harmonious totality caused by the cosmic energy is not experienced by human beings as a fortunate reality, but as a suffocating and oppressive ordeal that human intelligence—a privileged product of the cosmic vitality—must find a way to radically alter or escape from in a utopian flight.

Nahui Olin starts *Energía cósmica* by defining the "law of cosmic energy" ("ley de la energía cósmica") that provides the foundation for her cosmological speculations: "All atoms lack the will to direct their movement, their cosmic energy is their fundamental cause and their dissociation or their contact with another atom is involuntary [it is] only the impulse of total movement—the infinite that cannot degenerate into energy because its multiplicand is its dynamo and only that impulse of the cosmos, is the reason for all existing phenomena" ("Todos los átomos carecen de voluntad para dirigir su movimiento, su energía cósmica es su causa fundamental y su disociación o su contacto con otro átomo es involuntario sólo el impulso del movimiento total—el infinito que no puede degenerar en energía porque su multiplicando es su dínamo y sólo ese impulso del cosmos, es el porqué de todos los fenómenos existentes").[61] This definition already contains the main terms—"energy," "movement," "totality," "infinite," "phenomena"—that sketch out the picture of Nahui Olin's conception of the cosmos as described earlier. According to her, the universe is endowed with a vital quality that stems from the limitless diversity of interrelated entities and movements found everywhere. As she puts it:

> A movement in any direction that is of different trajectory and speed, develops in the smallest space that it occupies and in the time of its travel, a series of vibrations emanating from the radius that encompasses its strength, its shape and its march and are incalculable vibrations in their totality they become an infinity in diversity... in each infinitely small movement there is a multiplication to infinity, which is the vitality of the cosmos.
>
> *Un movimiento en cualquier sentido que sea de recorrido y rapidez diferente, desarrolla en el espacio más ínfimo que ocupa y en el tiempo de su recorrido, una serie de vibraciones emanantes del radio que abarca su fuerza, su forma y su marcha que son vibraciones incalculables en su totalidad se vuelven en diversidad un infinito... en cada movimiento infinitamente pequeño hay una multiplicación hasta el infinito que es la vitalidad del cosmos.*[62]

This cosmic vitality is underpinned by a network of relations in which each cosmic phenomenon impacts in countless and fluctuating ways the entities around it, which in turn affect other adjacent entities in a harmonious way, always maintaining the equilibrium of the universe conceived as a steady-state system: "The totality is a time, an endless elastic space where all the simultaneous movements produced independently of one another fit, all the relative spaces and times without altering the harmony of the powerful movement of the totality that encloses infinite movements of any kind always subject to its dominant force" ("la totalidad es un tiempo, un espacio elástico sin fin donde caben todos los movimientos simultáneos producidos independientemente unos de otros, todos los espacios y tiempos relativos sin alterar la armonía del movimiento poderoso de la totalidad que encierra infinitos movimientos de cualquier especie siempre supeditados a su fuerza dominante").[63] This infinite and harmonious totality produces and sustains every existing phenomenon in the universe, including human beings, while preserving its overall stability by functioning as a closed system: "We clearly see that nothing new leaves or enters it [the system], but with its own cosmic elements it maintains its inexhaustible vitality, hence the result of monotonous evolution and similar purposes" ("claramente vemos que nada sale ni entra nada nuevo en él [el sistema], sino con sus mismos elementos cósmicos mantiene su vitalidad inagotable, de ahí el resultado de evolución monótona y de finalidades semejantes").[64] The overall unchanging uniformity of the cosmos is experienced by human beings as a suffocating experience, since they are forcibly bound to living "always within the enormous and terrible totality that crushes us, nullifies us" ("siempre dentro de la enorme y terrible totalidad que nos aplasta, nos nulifica").[65] Human intelligence—being the product of cosmic energy concentration at the highest level—is capable not only of understanding the universe as a whole, but also of aspiring to transform it radically. This utopian drive is the expression of Nahui Olin's nonconformity regarding the monotonous nature of the universe: "The fact of breaking the infinity of related movements that form a totality, would satisfy my desire that is more infinite than infinity itself, since it finds it limited and determined to a cosmic production always similar in matters, species and phenomena" ("el hecho de romper la infinidad de movimientos relacionados que forman una totalidad, satisfacería mi deseo que es más infinito que el propio infinito, puesto que lo encuentra limitado y determinado a una producción cósmica siempre semejante en materias, especies y fenómenos").[66]

Thus, Nahui Olin puts forth the utopian desire to create an imbalance or destruction of the cosmic totality, which is paradoxically the ultimate manifestation of the vitality of the cosmos. But how can human beings achieve the ambitious and monumental purpose of destabilizing the cosmos? For starters, Nahui Olin is clear in arguing that scientific thinking and methods do not have the ability of accomplishing this grand undertaking: "Science is an encyclopedia of definite observations insufficient to find the indefinite atom that shall break the cosmic equilibrium through its paths, and I walked away from it because it was only a system of relative measures for possibilities of mediocre inventions, and I walked alone with the eyes of my intelligence, with the force of my reflection" ("La ciencia es una enciclopedia de observaciones definidas insuficientes para encontrar por sus caminos el átomo indefinido que ha de romper el equilibrio cósmico, y caminé lejos de ella porque sólo era un sistema de medidas relativas para posibilidades de inventos mediocres, y caminé sola con los ojos de mi inteligencia, con la fuerza de mi reflexión").[67] While Nahui Olin incorporates scientific terms in her work, she also suggests that relying strictly on science is ultimately an insufficient approach, because scientific thinking tends to paralyze and confine the infinite vitality of the cosmos. When scientists aspire to calculate the exact measures of the natural world, they only confine themselves to understanding one of the endless features of the universe: "they abstract an infinity from this point that they call determined with a number, a word, a formula" ("hacen abstracción de un infinito en este punto que ellos llaman determinado con un número, una palabras, una fórmula").[68] Although scientific calculations and measurements do achieve a utilitarian purpose by fostering technology, they are unable to engage with the limitless possibilities of the cosmos, and thus they become nothing more than a reiteration of the monotonous vicissitudes of the totality.

Like science, art is equally incapable of accomplishing the destruction of the universe's suffocating equilibrium. In spite of their own intentions, artists limit themselves to creating instances of "local artificial work and only for us because the Universe does not change at all with those improvisations of solely human character" ("obra local artificiosa y únicamente para nosotros porque el Universo no cambia en nada con esas improvisaciones de carácter solamente humano").[69] In other words, even when artists pride themselves on engaging in radically creative efforts, their acts of creation cannot achieve an absolute degree of inventiveness, to which Nahui Olin aspires. As she argues, "in works of art there is no creation, there is only a bad inanimate reproduction of existing colors and forms" ("en las obras de arte no hay ninguna creación, solo hay una mala reproducción inanimada de colores y de formas existentes").[70] Ultimately, artwork is just a different organization of the existing cosmic phenomena, and it is consequentially restrained within the boundaries of the universe. For example, painters solely translate their individual vitality in lines, colors, and figures in the canvas, which only represent a slightly diverse arrangement of the same cosmic energy: "The most valuable painters have always produced things that are within their own movement and the trend of their lines is nothing more than the movement itself that drives them to life and that movement is seen in the repetition of itself in the lines of such or such forms, within which they always hold their production" ("los pintores más valiosos han producido siempre cosas que están dentro de su propio movimiento y la tendencia de sus líneas no es más que el propio movimiento que los impulsa a la vida y ese movimiento se ve en la repetición de él mismo en las líneas de formas tales o cuales, dentro las cuales sujeta siempre su producción").[71] In sum, art does not have the ability of surpassing its limited area of impact and radically reinventing the fabric of the cosmos.

While science and art prove themselves to be fundamentally ineffective, Nahui Olin placed her hope in a direct, material transformation of the universe through the power of the human brain, which she called "cerebral dynamism" ("dinamismo cerebral"). Not unlike Dr. Atl, Nahui Olin believed in the possibility of a "connection between a micro and macrocosm" ("conexión entre un micro y macrocosmos") by establishing an unmediated interaction between human beings and the cosmic vitality: "when I have immediate contact in my body with nature and I focus on myself, I accumulate an amount of radio-activity and the radio-activity and the force that escapes from my pores begins to have an action on the immediate things that surround me, they feel me and I can possess them deeply in all their ignored dimensions" ("cuando tengo contacto inmediato en mi cuerpo con la naturaleza y me concentro conmigo misma, acumulo cantidad de radio-actividad y la radio-actividad y la fuerza que se escapan de mis poros comienza a tener acción en las cosas inmediatas que me rodean, me sienten y puedo poseerlas profundamente en todas sus dimensiones ignoradas").[72] Particularly, Nahui Olin was convinced of the infinite potentiality of the human brain, which—while being part of the general cosmic matter—possessed unique characteristics that made it stand out. Nahui Olin attributed a special kind of radioactivity to the brain to the extent that it was able to emit energy in the form of waves or vibrations. Thus, she set out to investigate and imagine the possibilities of the brain's radio-activity for materially interacting with the cosmic totality: "Every discovery in brain matter has always been discarded, it has been considered to be within the general mass—I do not divide it but I grant it a more important and energetic special radio-activity force to be the point that leads to a discovery of effective or positive value in the world" ("Siempre se ha desechado todo descubrimiento en la *materia cerebral*, se ha considerado dentro de la masa general—yo no la divido pero le concedo una fuerza de radio-actividad especial más importante y enérgica para ser el punto que lleve a un descubrimiento de valor efectivo o positivo en el mundo").[73] While art and science remain at the level of perception and representation, Nahui Olin accorded the human brain the potential of transforming the cosmos in a direct and material way. Consequentially, her utmost desire was "to intensify in the most powerful form of expression—in electrical vibrations—in brutal commotion over beings, transmitting absorbing atoms to the pores of their organism in this transformation of matter and forces, infusing into them the amazing force of cerebral dynamism that consumes matter and gets further development in what is called intelligence" ("intensificar en la más poderosa forma de expresión—en vibraciones eléctricas—en conmoción brutal sobre los seres, transmitiéndoles en esta transformación de materias y fuerzas, átomos absorbentes a los poros de su organismo, traduciéndoles la fuerza asombrosa del dinamismo cerebral que consume la materia y obtiene mayor desarrollo en lo llamado inteligencia").[74]

The idea of directly transmitting the cerebral energy underpinned several speculative proposals put forward by Nahui Olin. She referred to "the surely unconscious mechanics of this totality where we are included" ("la mecánica seguramente inconsciente de esta totalidad donde estamos comprendidos") in order to highlight how the movements of matter across the universe betray an oblivious and uniform mechanics that subsumes human beings under its relentless logic.[75] Most of humans tend to accept this "unconscious" mechanics without thinking they can do something about it. However, there exist human beings who develop an acute understanding of the cosmic vitality and become aware of their own reduced place in the totality. This segment of humans is no longer willing to be a trivial piece of the cosmic totality, and thus strive to impose their own consciousness on the universe as a whole: they aspire to "to invent a movement that is not adaptable to that unconscious mechanics and the maladjustment of which could produce a movement controlled by a conscious force, our brain that, harmonizing within that old unconscious mechanics, will be perfected in that new dissociation" ("inventar un movimiento inadaptable a esa mecánica inconsciente y que su inadaptación produjera un movimiento controlado por una fuerza consciente, nuestro cerebro que armonizando dentro de esa antigua mecánica inconsciente se perfeccionará en esa nueva disociación").[76] In other words, Nahui Olin envisioned that highly developed human brains could emit a material force that would deliberately introduce a kind of dissonant movement in the cosmos, creating an unbalance in the fabric of the totality: "I think about disintegrating all the harmony of movements of relation to impose the one that I bring, unique, multiplying myself and without anything else than me existing" ("pienso en desintegrar toda la armonía de movimientos de relación para imponer el que traigo único, multiplicándome yo misma y sin que nada más que yo existiera").[77] In the end, Nahui Olin's godlike ambition pushed her to desire a total destruction of the universe in order to establish her own boundless energy, becoming "an absolute thing of structure and thought where I exist in all the forms of my changing and perfect creation that would never have created such a nullifying thing as the existing cosmos" ("una cosa absoluta de estructura y pensamiento donde existo yo en toda la forma de mi creación cambiante y perfecta que nunca hubiera creado una cosa tan nulificante como el cosmos existente").[78] Similarly, in *Óptica cerebral* Nahui Olin provided an alternative image in which the human brain would also disintegrate the whole universe, but at that moment it would create a new cosmos with unprecedented vegetal and mineral kingdoms:

> Each spirit is a greater or lesser solar system, living in the only infinity ... and such is its evolutionary force that it would absorb everything in space, everything in what we call infinity, and as an immeasurable force of attraction it would paralyze all solar systems and absorb them into their matter as secondary juices for its exterior envelope, and from the mystery of infinity in the evolution of its spirit, only a star of indefinable matter would be born, because in the intense and continuous evolutionary force that maintains its existence, there would not be a thousandth of a second that would enter into the rapid transformation of its substances that, when enriched, would form unknown plants, minerals and gases.

> *cada espíritu es un sistema solar mayor o menor, que vive en el único infinito ... y tal es su fuerza evolutiva que llegaría a absorber todo lo que hay en el espacio, todo lo que hay en lo que llamamos infinito, y como una inconmensurable fuerza de atracción paralizaría todos los sistemas solares y los absorbería en su materia como jugos secundarios para su exterior envoltura, y del misterio del infinito en la evolución de su espíritu nacería sólo un astro de materia indefinible, porque en la intensa y continua fuerza evolutiva que mantiene su existencia, no habría milésimo de segundo que entrase en la rápida transformación de sus sustancias que enriquecidas, formarían vegetales, minerales y gases ignorados.*[79]

Nahui Olin proposed an additional way in which the "materia cerebral" could conceivably achieve the purpose of transforming the cosmic totality. She built on the idea that "the intra-atomic force of matter has been calculated ... and it is considered that the day a wise man finds the means of economically liberating the forces contained in matter, he would almost instantly change the face of the world" ("se ha calculado la fuerza intra-atómica de la materia ... y se considera que el día en que un sabio encuentre el medio de liberar económicamente las fuerzas que contiene la materia cambiaría casi instantáneamente la faz del mundo").[80] At the beginning of the twentieth century, the scientific research on the disintegration of elements conducted by physicist Ernest Rutherford opened up the prospect of harnessing the energy contained in atoms. H. G. Wells first imagined in his 1914 novel *The World Set Free* the idea that a massive amount of energy could be released by splitting an atom. By the 1930s, this prospect became a reality when scientists were able to employ nuclear fission as a weapon of mass destruction. Building on this research, Nahui Olin imagined that, given the fact that the human brain comprises the greatest concentration of energy in the universe, by splitting the atoms in cerebral matter one could release the most impressive quantity of energy ever recorded, transforming the cosmos in an instant: "it should be considered that the day when the means of economically releasing the forces contained in brain matter is found, it would instantly and wonderfully change the universe" ("se debería considerar que el día en que se encuentre el medio de liberar económicamente las fuerzas que contiene la materia cerebral cambiaría instantáneamente y maravillosamente el universo").[81] According to Nahui Olin, the fact that scientists have not looked into the possibility of releasing the energy stored in cerebral matter is a sign of modern science's powerlessness and unwillingness to embrace the most meaningful challenges. Accordingly, the project of "creating a new universe with the use of the power of brain radioactivity" ("crear un nuevo universo con el aprovechamiento de la fuerza de la radio-actividad cerebral") spearheaded Nahui Olin's intention to "create a new science, that of ourselves, from which new forces will emerge" ("crear una ciencia nueva, la de nosotros mismos de donde saldrán fuerzas nuevas").[82]

As part of this "ciencia nueva," Nahui Olin imagined one final utopian way in which the human brain could, as she puts it, "rebel against the systems that have turned us into atoms or forces as balancing pellets for a wheel in continuous motion" ("rebelarse contra los sistemas que nos han convertido en átomos o fuerza como balines de equilibrio para una rueda en movimiento continuo").[83] Cosmic energy is neither created nor destroyed and maintains a state of dynamic equilibrium on Earth. However, Nahui Olin suggested that the "dinamismo cerebral" could absorb and accumulate the energy from all entities in such a way that it would alter the equilibrium by reaching outer space:

> To bend, to overcome the dynamic energies of every being, of everything in its unconscious state, impotent to exceed that force, which bends them and uses them as one thousandth of one further impulse to the movement of their existence, make of all this an effort or an impossible step to get out of the atmosphere, which normalizes our forces to balance them in a group

called Earth, a planet where humanity exists, of potentiality X impossible to emerge beyond and become independent, since they are eternally attracted and weakened by the attraction of the earth.

Doblegar, superar a las energías dinámicas de todo ser, de toda cosa en su estado inconsciente impotente para sobresalir a esa fuerza, que los doblega y que los utiliza como un milésimo de impulso más al movimiento de su existencia, y hacer con todo esto un esfuerzo o un paso imposible para salir de la atmósfera, que normaliza nuestras fuerzas para equilibrarlas en un conjunto que se llama Tierra, planeta donde existe la humanidad, de potencia X imposibles de surgir más allá e independizarse, pues son eternamente atraídas y debilitadas por la atracción de la tierra.[84]

In other words, while the same flow of cosmic energy is usually captured and administered by the Earth, hampering any outflow beyond its frontiers, Nahui Olin thought that, by collecting the scattered energy, the human brain could escape from the control of the Earth's atmosphere.

Before humanity reached space in the 1960s, the theoretical possibility of venturing into space was studied by Russian cosmist Konstantin Tsiolkovsky, American physicist Robert H. Goddard and the Romanian-German rocket theorist Hermann Oberth during the first decades of the twentieth century. Of course, the topic of space exploration was also predominant in the science fiction work of figures such as H. G. Wells, Jules Verne, and Camille Flammarion. In line with this background, Nahui Olin's imagination—as well as Dr. Atl's, as I will show soon—was captured by the image of humanity's escape from the Earth toward outer space. Rather than focusing on the science or technology behind space exploration, she was particularly interested in how venturing into space opened up the possibility of radically transforming the cosmos. The act of "salir de la atmósfera" was marked by the utopian potential of fleeing from the cosmic totality and inaugurating a new state of being. According to her, reaching space was less a matter of constructing suitable rocket ships or any kind of technology, but more a quest that had to be undertaken by the human spirit and intelligence in a supreme act of rebellion. She acknowledged that it constituted an "impossible fact, incredible to be verified since they are unbreakable laws" ("hecho imposible, increíble de verificarse puesto que son leyes inquebrantables") but added that "the mere fact of feeling it, of having the absolute need for it, thinking about it means overcoming, leaving the universe" ("el solo hecho de sentirlo, de tener la necesidad absoluta de ello, el pensarlo significa sobreponerse, salir del universo").[85]

Nahui Olin developed her cosmological speculations during a time in which important discoveries and theories were changing the face of scientific cosmology and establishing a new conception of the universe. Since the times of ancient Greek philosophy, cosmological thinking had pondered two central issues: the question of whether the universe has always existed or was originated at a specific point in time; and the question of whether the universe has always maintained a fundamentally stable state or has endured a process of evolution with important transitions. During the age of Enlightenment, the notion of a stable and infinite universe governed by natural laws became generally accepted by astronomers and philosophers. But this consensus became less secure in the first decades of the twentieth century, when Albert Einstein's general theory of relativity of 1915 applied to cosmology by Georges Lemaître, along with Edwin Hubble's groundbreaking astronomical observations in the 1920s provided the foundation for a new conception of the cosmos.[86] During this period, two basic notions of modern cosmology were formulated for the first time: the notion that the universe had an explosive origin at a determined point in the past (later known as the theory of the Big Bang), and the proposition that the universe is expanding continually and thus evolving gradually. Even though these two notions were hotly contested at the time, by the 1950s they eventually formed the general consensus approved by cosmologists.

As analyzed earlier, Nahui Olin accepted a conception of the universe as a dynamic yet unchanging, infinite totality, powered by a constant flow of cosmic energy. This conception is clearly in consonance with the traditional image of the universe and thus seemed to disregard—as most scientists still did at the time—the important cosmological theories and discoveries of the 1920s and 1930s. In fact, in *Energía cósmica* Nahui Olin explicitly criticized Einstein's theory of relativity as "a human formula, an isolated assumption, but it is not a general law of the cosmos" ("una fórmula humana, una suposición aislada, pero no es una ley general del cosmos"), because in the end it constituted a reductionist approach that attempted to collapse the endless nature of the universe into a necessarily limited, human view.[87] "Einstein," as Nahui Olin puts it "is admirable because he has discovered the law of relativity, time and space, but how has the universe changed with that?—In nothing" ("Einstein es admirable porque ha descubierto la ley de la relatividad, tiempo y espacio, pero ¿en qué ha cambiado el universo con eso?—en nada").[88] While Einstein's theory—and science or art in general— were ultimately ineffective, Nahui Olin aspired to materially transform the cosmic totality by either accumulating cosmic energy and inventing a disruptive movement, or by releasing the energy stored in the human brain, or even by escaping from the Earth's atmosphere. Nahui Olin expressed thus a radical discontent with the traditional view of the cosmos as a stable and infinite totality, and believed that the purpose of human beings was to embrace the cosmic vitality and create a new universe. In this sense, even if she did not incorporate the novel theories and discoveries of her time, Nahui Olin's cosmological thought manifested in its own way the general intention of finding alternatives to the long-standing, static view of the cosmos. As I will show next, Dr. Atl's cosmological speculations similarly aimed to surpass the limits of human knowledge on cosmology by physically exploring the frontiers of the universe.

BEYOND THE UNIVERSE

Dr. Atl's *Un hombre más allá del universo* (1935) is a science fiction novel that recounts a man's journey to discover the outer limits of the universe.[89] Not unlike Nahui Olin's view, this novel also suggests that insofar as science is incapable of establishing a material relation with the universe, human beings must introduce a direct, physical interaction with the cosmos by exploring beyond Earth's atmosphere. Space exploration is construed as a utopian undertaking that provides close knowledge of the universe unmediated by science's limited and inadequate approach. The novel shows that Dr. Atl was cognizant of recent cosmological scientific theories and sought to engage speculatively in the ongoing conversation on contentious issues such as the shape, structure, and functioning of the cosmos, or the prospect of a universal center. However, more than a method of acquiring knowledge, space exploration had for Dr. Atl the potential of recreating the cosmos in the process of traversing it. In consonance with Nahui Olin's thought, Dr. Atl's novel reveals that humanity possesses the ability to directly tap into cosmological forces to control and reorganize the universe according to its needs. At the same time, the force that drives the protagonist's journey in the novel is the idea of arriving at the "más allá" of the universe, that is, encountering the limits of the cosmos and all human knowledge. Paradoxically, this utopian promise of the "más allá" challenges all conceived notions of the cosmos, of space and time, and thus undermines the protagonist's own intention of exerting control over the universe.

Unlike Dr. Atl's visual artwork, *Un hombre más allá del universo* has received little critical attention probably due to its radical departure from the predominant literary trends of nineteenth-century and postrevolutionary Mexico. In the 1980s, Françoise Perus argued in a critical reading that this novel showcases Dr. Atl's monist and mechanistic understanding of the universe. According to Perus, this conceptualization amounts to a rather ingenuous and undialectical philosophical materialism that betrays Dr. Atl's "impossibility of thinking history and of thinking himself in history" ("imposibilidad de pensar la historia y de pensarse a sí mismo en la historia").[90] More recently, Cuauhtémoc Medina and Renato González Mello have proposed readings that incorporate an examination of the novel's engagement with science. While González Mello examines the novel's affinities to Alfonso L. Herrera's ideas, to which I will come back later in my analysis, Medina weaves together textual analysis with a consideration of Dr. Atl's unpublished papers in order to maintain that this novel "divulges, surreptitiously, a new conception of the Cosmos and of men within it" ("difunde, subrepticiamente, una nueva concepción del Cosmos y del hombre en su seno").[91] According to Medina, this conception has a clear affinity with esoteric ideas, and ultimately relies on a nihilist and post-theological approach—inspired by Nietzsche's thought—that calls for a refoundation of the cosmos. Medina also demonstrates that *Un hombre más allá del universo* presents in a narrative manner Dr. Atl's "ciencia alternativa": his ideas and speculations concerning physics and the cosmos that he developed in an extensive assortment of unpublished papers. However, Medina does not incorporate an examination of the novel's intervention in cosmological debates of its time and how Dr. Atl endeavored to surpass them in search of transforming the universe.

Un hombre más allá del universo begins with a scene that resonates with *Las sinfonías del Popocatépetl* and remarkably ties together Dr. Atl's interest in volcanism and his attraction to space exploration. The narrative voice describes the protagonist climbing a volcano with the aim of reaching the peak and set up a camp there: "he was going to live on the abyss, detached from all fortuitous things in the world, sure of concentrating theories and experiments on the possibilities of traveling through internebular spaces—supreme ambition of his spirit—supreme human ambition: to conquer the Universe" ("iba a vivir sobre el abismo, desprendido de todas las cosas fortuitas del mundo, seguro de reconcentrar las teorías y los experimentos sobre las posibilidades de recorrer los espacios internebulares—suprema ambición de su espíritu—suprema ambición humana: conquistar el Universo").[92] As in Dr. Atl's writings on volcanism, here the volcano embodies the place that instantly puts the subject's body in the position of interacting intensely both with geological and cosmological forces. The volcano's crater is thus the perfect location for the protagonist's plans of producing a spaceship capable of "moving around the mysterious beyond" ("desplazarse en el más allá misterioso").[93] As he explains to the amazed visitors once his plans have been completed, the protagonist did not *build* this spaceship by mechanical or technological means. Instead, he condensed and organized with the power of his brain the cosmic energies summoned by the volcano's crater: "I have not done nothing more than establishing the points of relationship for Nature to operate by concentrating multiple energies, following the call of human power" ("yo no he hecho más que establecer los puntos de relación para que la Naturaleza operase reconcentrando al llamado del poder humano, múltiples energías").[94] In other words, just as Nahui Olin's approach to the "materia cerebral," the protagonist of Dr. Atl's novel has directly harnessed cosmological forces in order to fabricate a spaceship named "Cristal Cósmico," which possess properties—such as its ultra-resistant material, its perfect polyhedron shape, its surrounding atmosphere—that are unobtainable by any known scientific or technological techniques.

The protagonist goes on to explain that all scientific cosmological theories are ultimately inconsequential endeavors that contribute little to "carry out my program of material knowledge of the Cosmos" ("llevar a cabo mi programa de conocimiento material del Cosmos").[95] He makes an analogy between the world of fashion and cosmology: just as there are several fashion houses that strive to establish a unanimous trend, there are also "houses specialized in cosmogonic fashion" ("casas especialistas en modas cosmogónicas") including "famous houses such as Abbe Lemaitre's, Professor Jean's, Haeisenberg's, Dr. Shapley's, and Einstein's, which imposed on the clientele the style of their creations" ("casas famosas como la del abate Lemaitre, la del professor Jean, la de Haeisenberg, la del Dr. Shapley, la de Einstein, que cualquiera que sea el estilo de sus creaciones, se impone a la clientela").[96] All these "ultraterrestrial fashion houses are dressing the Universe with a new suit in each season, like the [fashion houses] of Paris do to women" ("casas de modas ultraterrestres, están vistiendo al Universo con un nuevo traje en cada estación, como las [casas de moda] de París a las mujeres").[97] According to the protagonist, both fashion and cosmological trends are ephemeral and superficial tendencies that are unable to make any substantive changes in the world. Cosmological trends aspire to measure and contain the whole universe in a single mathematical formula or figure, but "all these figures put on paper by my wise teachers, they neither measure anything nor mean anything" ("todas estas cifras puestas sobre un papel por mis maestros los sabios, ni miden nada ni significan nada").[98] All cosmological theories are bound to set a meaningless limit on the universe and remain silent on the most transcendent question of all: what is "beyond" the universe and "beyond" all human knowledge. In this sense, cosmological theories betray "the lack of adequate mechanisms to establish material relationships with the Universe" ("la falta de mecanismos adecuados para establecer relaciones materiales con el Universo"), an issue that will be finally addressed by the protagonist's exploration of the universe.[99]

By challenging all cosmological theories of the time, the novel's protagonist sets out to intervene in the most decisive debates in scientific cosmology and, eventually, transcend them in an attempt to materially recreate the cosmos. Some of the most controversial issues in cosmology came into play in the so-called "Great Debate" that took place in 1920 between Harlow Shapley and H.D. Curtis concerning the scale and structure of the Milky Way galaxy and the possibility of the existence of other galaxies outside our own. Up until the 1920s, there was considerable debate on whether the Milky Way galaxy encompassed the whole universe or just a part of it, whether other "nebulae"—now known as "galaxies"—existed outside of the Milky Way, and whether the solar system in which we live was located near or far from the center of the universe. In the so-called "Great Debate," two opposite views surrounding these issues were propounded: Shapley argued that the Milky Way galaxy had a previously unimaginable massive scale containing all other "nebulae," and maintained that our solar system was not located in the galactic center but many thousands of light years away from it, near the Sagittarius constellation; for his part, Curtis advocated the so-called "island universes theory," that is, the idea that there exist galaxies outside our own that are comparable in size to the Milky Way, and accepted that our solar system was located near the center of the galaxy.[100] In the next few decades, it became clear that both views were partially correct: while the solar system is roughly located where Shapley proposed and the scale of the Milky Way is close to his estimations, Curtis's premise concerning the "island universes" also became fully accepted. However, at the time of the debate, these were highly contentious issues that puzzled the imagination of cosmologists, philosophers, and artists, including Dr. Atl's cosmological novel.

In fact, after an initial description of the origin and properties of the protagonist's spaceship, the rest of *Un hombre más allá del universo* consists of a first-person account of his journey throughout the universe, implicitly touching upon some of the questions covered in the "Great Debate." At some point in his travel, the protagonist reaches the outer limits of the Milky Way and feels how the galaxy's rotation propels him into a space "out of the action and meaning of our stellar Universe" ("fuera de la acción y del sentido de nuestro Universo estelar").[101] He then moves across a string of galaxies that seems endless and uniform: "Nebulae follow one another, they multiply indefinitely. I touch one after another—they all seem alike to me—but are they really? Is there not a fundamental incapacity in me that prevents me from comparing them properly?" ("Las nebulosas se suceden, se multiplican indefinidamente. Toco una y otra—todas me parecen semejantes—pero lo son realmente? No habrá en mí una incapacidad fundamental que me impida comparar debidamente?").[102] While some "nebulae," or galaxies, seem to be comparable in size and shape, the protagonist eventually encounters one that functions in the exact opposite way from "our Universe," the Milky Way. This opposite universe not only displays an "inverted mechanics" ("mecánica invertida") in which the generating core lays outside and "its generating force radiates from the outside" ("su fuerza generadora irradia del exterior"), it also presents itself as a place where time passes in reverse: "What in our Universe is to come, here it has already happened, that is, here it's in the past" ("Lo que en nuestro Universo está por venir aquí ya pasó, es decir aquí es el pasado").[103] In this way, Dr. Atl's novel embraces the "island universes" proposal and speculatively suggests that some of these unknown "universes" may challenge our understanding of the whole cosmos.

Similarly, *Un hombre más allá del universo* engages in the debate concerning the center of the galaxy and its fundamental characteristics.[104] Even when some cosmologists and astronomers accepted that the solar system was not located in the galactic center, what kind of entity occupied that place—now known to be a massive black hole—remained one of the most perplexing questions in cosmology until the 1960s.[105] In Dr. Atl's novel, the protagonist reaches the galactic center and finds an indescribable scenario marked by a diversity of shapes, forces, and colors: "Blue waves—unknown force—imponderable radiations and somewhere, fixed, immense prisms radiate—glare—splendor—unthinkable colorations—curvilinear light—silent trepidation that creates spatial rotations—something like crystalline waves, fluid, blue, purple, with the irises of all the suns and the lights of all the stars" ("Ondas azules—fuerza desconocida—radiaciones imponderables y en alguna parte, fijos, inmensos prismas irradian—fulgor—esplendor—coloraciones impensadas—luz curvilínea—trepidación silenciosa que crea las rotaciones espaciales—algo como ondas cristalinas, fluidas, azules, violetas, con los iris de todos los soles y las luces de todas las estrellas").[106] From this unimaginable vision, expressed in terms akin to avant-garde poetry, the protagonist gathers that he is witnessing "an invisible center that can be felt but cannot be seen" ("un centro invisible que se siente pero no se ve"), the "immense source of universal matter" ("fuente inmensa de la materia universal"), a "immobile center—distributive stillness of universal life" ("centro inmóvil—quietud distributiva de la vida universal").[107] In other words, he has encountered the center from which emanates the unique flow of matter and energies that propel the galaxy's rotation and create the diverse forms of universal life. Dr. Atl believed that his representation of the galactic center had anticipated some of the most recent scientific findings in this field: in a 1940 letter to his friend Joaquín Gallo, director of the Observatorio Astronómico at Tacubaya, he even argued that Shapley's conception of a massive universe rotating around a galactic center was first enunciated by himself in *Un hombre más allá del universo*.[108] Even if Dr. Atl's cosmological assertions are not necessarily always accurate, what is unquestionable is that his imagination tackled through creative means some of the most formidable problems in cosmology.

Dr. Atl's conception of the universe still relies on a theory that was widely discredited in the scientific community of the 1930s: the alleged existence of an endless medium called ether that allowed the interaction of energy and matter. As Dr. Atl puts it in the novel, "The ether is an electrical fluid in motion that fills the entire universal cell and within which all existing things move" ("El éter es fluido eléctrico en movimiento que llena toda la célula universal y dentro de la cual se mueven todas las cosas existentes").[109] As González Mello has pointed out, Dr. Atl's strict reliance on the theory of ether was also shared by Herrera and it served as the basis of the materialist, monist philosophy espoused by both intellectuals: "To him [Dr. Atl] and to Herrera the ether was important because both of them, by different paths, had decided that it was not possible to sustain a metaphysical principle for life. The ether united the living with the nonliving, the planet with life around it, the earth with the galaxies" ("Para él [Dr. Atl] y para Herrera el éter era importante porque los dos, por caminos distintos, habían decidido que no era posible sostener un principio metafísico para la vida. El éter unía a los vivos con los nos vivos, al planeta con la vida a su alrededor, a la tierra con las galaxias").[110] In other words, the theory of ether afforded a non-metaphysical explanation for the existence of universal life, while also providing a unifying principle behind all existing phenomena. This "ethereal philosophy"—developed by Herrera in his philosophical writings, as I have showed in Chapter 2—thus underpins the search of the novel's protagonist to discover the limits of the universe.

Further along his journey, the protagonist has the opportunity of observing the diversity of life forms created by the universal flows of matter and energies emanating from the galactic center. First, he encounters a planet composed entirely of gold, inhabited by winged monsters of dimensions so massive that "each one of its flaps could extinguish our sun" ("cada uno de sus aletazos podría apagar nuestro sol").[111] While these monsters are brute and violent, the beings that the protagonist meets later in his journey are infinitely more intelligent and sophisticated than humanity. These beings instantly communicate without words with the protagonist and let him know that, despite their apparent differences, they, like humans, are also a product of the same cosmic energies that populate the whole universe. Among all the forms of universal life, these highly intelligent beings are special because they have been able to harmoniously rearrange their galaxy by linking all of its parts with luminous threads made of electricity that create a marvelous fabric. The protagonist is amazed at how these superior beings "have been able to make of a whole nebular system, what we do on Earth between two devices with an electric current" ("han podido hacer de todo un sistema nebular, lo que nosotros hacemos en la Tierra entre dos aparatos con una corriente eléctrica"), demonstrating that "only intelligence is capable of disorganizing the Universe—and also of organizing it artificially" ("sólo la inteligencia es capaz de desorganizar el Universo—y también de organizarlo artificialmente").[112] Whereas the act of thinking is usually construed as an "internal" process on Earth, in this planet the intelligence "is an undulation that is externalized" ("es ondulación que se exterioriza") and has an immediate effect on the universe.[113] These beings provide the protagonist with a living example of how humanity can control the universe as a whole by harnessing cosmological energies through the application of intelligence's material power directly to the cosmos. In other words, these intelligent beings have mastered and extended the procedure that the protagonist himself followed to produce his spaceship on Earth and that humanity must follow in the future to reach new levels of perfection and power.

Even though the protagonist praises these highly intelligent beings, he makes a conscious decision to leave this planet and continue his expedition across the universe. As he tells them, "I admire your perfection and your power and the rhythm of your life, but I prefer human imperfection and human passion and unsatisfied desire and fleeing life and death that approximates" ("admiro vuestra perfección y vuestra potencia y el ritmo de vuestra vida, pero prefiero la imperfección humana y la pasión humana y el deseo insatisfecho y la vida huyente y la muerte que llega").[114] In other words, although the protagonist has seen firsthand how humans could exert absolute control over the cosmos and adjust it to their needs, he decides to embrace humanity's persistent longing for new realities.[115] Thus, there is a constitutive tension in the novel between the aspiration to "conquer" the universe through the power of intelligence, and the desire to challenge all human ambitions of control and supremacy over the cosmos. Ultimately, the protagonist's journey is driven simultaneously by the intention of dominating the universe by understanding its functioning, and by the inexhaustible drive to reach the "más allá": the utopian place in which all human knowledge and authority stop making sense.[116] In resuming his journey through the cosmos, the protagonist is fully aware that reaching the limits of the universe and human knowledge is the fuel that keeps him going forward: "If I am not to reach the unknown beyond even of your own wisdom, the beyond of these spaces limited by an impenetrable darkness, I prefer to stop at the blackest point of the unknown, exit my crystal and throw myself naked into the void!" ("Si no he de alcanzar el más allá desconocido hasta de vuestra propia sabiduría, el más allá de estos espacios limitados por un oscuridad impenetrable, prefiero detenerme en el punto más negro del ignoto, salir de mi cristal y lanzarme desnudo en el vacío!")[117]

The aforementioned tension reaches its culminating point and seems to be resolved at the end of the novel. After a long journey full of unexpected adventures, the protagonist eventually encounters "something that is not a succession of events like the Universe, nor a symbol like Divinity, nor a vibration like light, nor a series of ever-changing and evolving phenomena like life" ("algo que no es una sucesión de eventos como el Universo, ni un símbolo como la Divinidad, ni una vibración como la luz, ni una serie de fenómenos cambiantes y en continuo devenir como la vida").[118] This mysterious "absolute entity" exceeds all human paradigms by undermining customary notions of space, time, shape, and sound: it draws a perfect straight line that appears to destroy the universe's curved space, while also producing a solidifying sound that starts to absorb everything around it. In the face of this entity, which embodies the limits of the universe and human knowledge, the protagonist feels utterly lost and confused: "I don't know if I move my arms in immeasurable planes and if my eyes perceive what does not exist, or if the incomprehensible has really absorbed me, or if I am absorbing the incomprehensible" ("No sé si muevo los brazos en planos inconmensurables y si mis ojos perciben lo que no existe, o si realmente lo incomprensible me ha absorbido, o si estoy absorbiendo lo incomprensible").[119] The protagonist has apparently achieved his purpose of reaching the "más allá" of the universe, but this experience ends up filling him with an existential dread that paralyzes his body and seems to anticipate the end of the cosmos. Suddenly, at the most dramatic moment, the protagonist feels that unimaginable flows and energies abruptly exit his body and brain. Without warning he experiences the most transcendental event: "A Universe is born! A spark blazes into nothingness—a spark that breaks the unknown, a blaze grows, grows, and spirals, saturated with nebulae. Another Universe arises, all of a sudden, in an instant, complete, with its suns and its worlds, and there I am again within it, living in a molecule saturated with hope!" ("Un Universo nace! Un chispazo fulgura en la nada— una chispa que rompe lo desconocido, un fulgor crece, crece y se mueve en espiral, saturado de nebulosas. Otro Universo surge, de un golpe, en un instante, completo, con sus soles y sus mundos, y ahí estoy yo otra vez dentro de él, viviente en una molécula saturada de esperanza!")[120]

By reaching the limits of the universe and of knowledge, the protagonist has destroyed the existing universe only to create a new one. In contrast with Cuahutémoc Medina's reading that characterizes this ending as a deification of the protagonist, I argue that this supremely creative outcome—which resonates with Nahui Olin's suggestions—is the unpredictable product of a line of flight that destabilizes humanity's aspirations to conquer and govern the universe.[121] In other words, the protagonist has employed the power of intelligence to a point where he, more than governing the current universe as the highly intelligent beings did, has originated the potentiality for a new terrain and a novel set of rules. This, in turn, has introduced simultaneously a new set of limits and a new sense of "más allá" that the protagonist aspires to reach once again. Thus, the novel suggests that the infinite power of the brain resides not in administering the universe—which is, after all, science's limited approach—but in engaging in a continual, always unfulfilled, utopian desire to search for the new beyond the universe. As I will examine next, Dr. Atl's cosmological novel resonates with another utopian undertaking that would captivate his imagination and dominate the last years of his life.

OLINKA

Dr. Atl's most transcendent project, to which he attributed the greatest importance for the future of humanity and the universe itself, was the construction of an ideal city named Olinka, meaning "where the movement is concentrated" ("donde se reconcentra el movimiento") in Nahuatl, a place where the most prominent minds of the world would gather together to investigate and accomplish the next stage of humanity's evolution. As Cuauhtémoc Medina has extensively shown, the project of building an ideal city went through several iterations since the early 1910s until Dr. Atl's death in 1964, with a period of intense activity in the 1950s.[122] Over the course of the decades, the project changed substantively in terms of the city's construction site, its primary goals, its architectural and administrative design. But overall, this utopian project condensed Dr. Atl's interests in cosmological and geological issues, as well as his fascination with the unfulfilled potential of the human brain, with the aim of charting the way for humanity's progression in the universe.

The first mention of the project of constructing an ideal city appears in a 1912 letter to Federico Gamboa, in which Dr. Atl speaks of his intention of "to found a comprehensive city, where men of all races, (under) the infinite shelter of Beauty, could unleash all the energies that are suppressed or diverted by the currents of needs of our social state" ("fundar una ciudad integral, donde los hombres de todas la razas, (al) amparo infinito de la Belleza, pudieran dar rienda suelta a todas las energías que suprimen o desvían las corrientes de necesidades de nuestro estado social").[123] In 1913, while living in Paris and participating in the avant-garde group Action d'Art, Dr. Atl undertook his initial attempts to make his dream into reality. According to his own testimony, he was able to secure a large extension of land south of Paris and two bankers committed to helping him raise money, but at that time he received an invitation to return to Mexico and joined the armed revolution.[124] As argued by Medina, this first iteration of the ideal city was heavily influenced by Dr. Atl's engagement with the Action d'Art group, which sought to question the role of art in capitalist society and assert the intellectual superiority of artists.[125] Accordingly, the 1910s proposal of the still unnamed city placed special emphasis on aesthetic concerns and made no mention of cosmological ideas.

After his return to Mexico, Dr. Atl seemingly put aside his city project until the 1940s, when he restated in an unpublished manuscript the utmost importance of building a city that could bring together the foremost minds of the world in one place. This manuscript called "Artistocracia"—a word play with "artista" and "aristocracia"—reaffirmed the aesthetic rationale behind the project, but also included for the first time an allusion of the fundamental role that natural sciences—in particular, astronomy—would play in the future city.[126] The cosmological underpinning of the project finally came to the fore in the 1950s along with Dr. Atl's most active commitment to accomplishing his plans. In 1952, he founded an organization that gathered renowned scientists, artists, and politicians with the aim of spearheading his efforts of securing the necessary means to construct the ideal city in Mexico. The organization published a pamphlet called *Crear la fuerza* in which Dr. Atl argued that science—not art as he had maintained in the 1910s—was the privileged domain for bringing the radical transformation of humanity to fruition: "The maximum activity of modern man is engendered by science. I mean that humans have created a new force to open the routes of the future. These routes need to be expanded and prolonged. A directing center, *a center for intellectual planning*, is necessary to guide evolution toward a new goal. That new goal is the real conquest of the Universe" ("La actividad máxima del hombre actual está engendrada por la ciencia. Quiero decir que el hombre ha creado una nueva fuerza para abrir las rutas del futuro. Es necesario que esas rutas se amplíen y se prolonguen. Un centro director, *un centro de planificación intelectual* es necesario para encauzar la evolución hacia una meta nueva. Esa meta nueva es la conquista real del Universo").[127] Thus, this pamphlet reveals how the aesthetic objective of the 1910s project shifted from the 1950s onward to the scientific ambition of finding the best way to travel to space and "to take the stars with your hand, step on them, measure the distances with our own body moving in space; to know *who* lives in other worlds, conquer them" ("Coger los astros con la mano, pisarlos, medir las distancias con nuestro propio cuerpo desplazándose en el espacio; saber *quiénes* viven en otros mundos, conquistarlos").[128]

In this pamphlet, as well as in other unpublished manuscripts such as *El futuro de la especie*, there is a tension similar to the one at play in Dr. Atl's science-fiction novel. On the one hand, he argues that space exploration constitutes the next stage in humanity's natural progression toward increasing its control over nature. He explicitly establishes an analogy between the conquest of space and the conquest of the American continent, suggesting that space travel will restart the will to power embodied by European conquistadors in the past: space exploration "signals a new course for the human species: the conquest of the unknown Cosmos, as the unknown land awakened the spirit of conquest in the past" ("señala un nuevo rumbo a la especie humana: la conquista del Cosmos desconocido, como la tierra desconocida despertó el espíritu de conquista en el pasado").[129] Just as conquistadors conquered "unexplored" territories along with their inhabitants, humanity must conquer other planets and everything else in the universe in the future. Yet this ambition to command the cosmos is contradicted by Dr. Atl's own suggestion that space exploration represents the best way to "direct it [the human will] toward a point, toward an ideal, toward a grandiose creation from which a completely new fact can arise" ("dirigirla [la voluntad humana] hacia un punto, hacia un ideal, hacia una creación grandiosa de la que pueda surgir un hecho completamente nuevo").[130] As Dr. Atl argues elsewhere, while every other realm of human activity has historical precedents that determined it in some way or another, space exploration constitutes a radically novel undertaking based on equally innovative scientific findings, which aim to "to trace a spiral from Earth into the unknown—a spiral neither mystical nor utopian nor philosophical, but real, physical" ("trazar una espiral desde la Tierra hacia lo desconocido—una espiral ni mística ni utópica ni filosófica, sino real, física").[131] In sum, Dr. Atl suggests that space exploration is both an unprecedented search for the unknown and at the same time the replication of the goal that guided the Conquest of Americas: controlling and administrating increasingly more territory and peoples.

As I argued earlier, this contradiction in Dr. Atl's work appears to resolve itself in *Un hombre más allá del universo* in favor of the infinite quest for novel, creative outcomes. In contrast, in the 1950s project of Olinka the will to increase human power and control over the universe seems to overtake the permanent search for the new. I contend that historical background accounts for this crucial difference between Dr. Atl's approach to space exploration in the 1930s and his perspective on this issue two decades later. In the 1930s, references to space travel in Dr. Atl's and Nahui Olin's work understandably displayed a rather speculative nature, because the science behind space exploration was in its initial stages and the outlook of surveying space seemed remote. In this context, space travel sparked all sorts of imaginative notions concerning the plausibility and implications of reaching the stars. From the 1950s onward, however, space exploration increasingly became a tangible possibility, not just an issue of utopian imagination anymore, as scientific breakthroughs provided the foundation for making it a reality. In 1957, a few years after Dr. Atl published *Crear la fuerza*, the Soviet Union successfully launched the first artificial satellite into orbit, ushering the space age and the start of the US–USSR space race that would culminate with American spacecrafts landing in the moon in 1969. Under these new circumstances, Dr. Atl's long-standing interest in space exploration took a more practical and authoritative tone, suggesting that his project of conquering the cosmos "is not the result of a desire for profit or of a more or less fantastic lucubrations, but the result of a contemporary, latent need" ("no es el resultante de un deseo de lucro ni de lucubraciones más o menos fantásticas, sino el resultante de una necesidad contemporánea, latente").[132]

Even though Dr. Atl understood that historical circumstances tended to legitimize his plans of reaching the stars, at the same time he was dissatisfied with the ways in which nations were attempting to accomplish this goal. First of all, he profoundly regretted that the space race was becoming a political contest between major world powers, instead of envisioning it as a joint task that demanded the best collaborative efforts of all nations, since it would ultimately determine the future of the human species as a whole. Dr. Atl projected that Olinka could bring together the most qualified scientists of the world, who would work collectively to design the best course of action to achieve the conquest of the universe. Undoubtedly, Dr. Atl considered that the current technological means employed by modern nations to travel to space—such as jet rockets, for example—were completely unsatisfactory, because they relied upon a limited scientific understanding of human potentialities. Particularly, all existing technologies disregarded the infinite potential of the brain and its power to directly harness cosmological forces, which—as I have shown before—was an issue explored in *Un hombre más allá del universo*. As Medina demonstrates, Dr. Atl genuinely believed that the brain's control of the universe unmediated by perception or representation was a promising scientific endeavor and he set out to develop it himself under the name of *cerebrología* in a cluster of works that remain unpublished to this day.[133] In some 1950s iterations of Olinka the goal of improving *cerebrología* became a central aspiration of the city along with space exploration: "my project inaugurates the construction of a true supercity in which most of the work that is being carried out in the world will be concentrated to create the new science of astronautics and, at the same time, another new science, cerebrology, that will be developed within a special institute" ("mi proyecto inicia la erección de una verdadera superciudad en la que van a reconcentrarse la mayor parte de los trabajos que se están realizando en el mundo para crear la nueva ciencia de la astronáutica y, al mismo tiempo, otra ciencia nueva, la *cerebrología*, que se desarrollará dentro de un instituto especial").[134]

Besides *cerebrología* and space exploration, geological phenomena also played an essential role in Dr. Atl's plans of constructing Olinka. All places that he deemed as potential construction sites throughout the years—the suburbs of Paris, between the Popocatépetl and Iztaccíhuatl volcanoes, Montebello lakes in Chiapas, Pihuamo valley in Jalisco, the Santa Catarina mountain range near Mexico City, the town of Tepoztlán in Morelos—were characterized by the presence of magnificent, untainted natural surroundings, some of them including volcanic structures and other geological phenomena. Particularly, Dr. Atl directed his best efforts to securing land near Santa Catarina in the 1950s, because he considered that this mountain range "is extremely steep and offers as a whole one of the most beautiful places in the Valley of Mexico, which remain unknown to date, despite its proximity to the capital" ("es extremadamente abrupta y ofrece en su conjunto uno de los lugares más bellos del Valle de México, hasta la fecha desconocido, a pesar de su cercanía a la capital").[135] His architectural ideas were based on incorporating the geological characteristics into the city's construction plan. First, he envisioned making use of La Caldera's volcano crater to build a massive sports stadium with a capacity of 300,000.[136] Apparently, Dr. Atl thought this plan was not as impressive or splendid as Olinka would need it to be, because a few years later he returned to this issue, suggesting that La Caldera's crater would be the most important place in the city: "This crater, which is surely the oldest volcanic formation in the valley, constitutes a magnificent geological structure. . . . It is not possible to establish, at the moment, a building program, but it is possible to imagine the possibilities of turning it into a place that can become, in all respects, the core of OLINKA" ("Este cráter, que es seguramente la formación volcánica más antigua del valle, constituye una magnífica estructura geológica. . . . No se puede establecer, por el momento, un programa de edificación, pero sí es posible imaginarse las posibilidades de convertirlo en un lugar que pueda devenir, bajo todos los aspectos, en el núcleo de OLINKA").[137]

Dr. Atl never provided a more detailed plan of how the city would take advantage of this volcano, but it is clear that he thought that this particular site was an essential factor—not just a random or meaningless component—for achieving Olinka's ultimate goal. Taking into consideration Dr. Atl's approach to volcanism, as well as its relation to cosmological ideas, it becomes easier to surmise the reasons for this. As analyzed earlier, Dr. Atl's writings on volcanoes and even his science-fiction novel build on the notion that engagement with volcanic structures opens up a material, bodily relationship between humans, the Earth, and the universe as a whole. Volcanic craters are privileged places that condense a comprehensive experience of the universe: this is, after all, the spot where the protagonist of *Un hombre más allá del universo* summoned cosmological energies in order to produce his spacecraft and conquer the cosmos. Thus, it is hardly surprising that Dr. Atl insisted on constructing the ideal city on and around a volcanic crater, because in the end he believed that it would contribute to further developing *cerebrología* and the technology for space exploration. In fact, a few years before drafting the proposal for constructing Olinka in Santa Catarina in 1954, Dr. Atl published his work on the Paricutín volcano in which he suggests the essential interrelation between volcanic and cosmological processes.

Medina argues that Dr. Atl's engagement with anti-Semitic Nazism and its eventual defeat at the end of the Second World war triggered his withdrawal from politics to restart his project of the ideal city, while surreptitiously adapting his ideas of the importance of an elite of superior men to the attainment of his cosmological dreams.[138] In this vein, one could also argue that Dr. Atl's desire to enhance elite power and control over the world first led to his support of Nazism, but after Hitler's resounding defeat this desire became a fundamental motivation that pushed him to revisit his project of an ideal city. As I have examined, Dr. Atl's long-standing engagement with cosmological and geological concerns, as well as his obsession with the infinite potential of the human brain, also converged distinctly in his dreams of conquering the universe by bringing together the foremost scientists in a splendid city. Thus, in the project of Olinka, Nazi will to power seemed to intensify and turn into the undertaking of controlling the whole cosmos.

Clearly, Olinka was not aimed at providing an ideal model of a national political community—a model of how Mexico or any other country should be structured to achieve general wellbeing—but instead it focused on proclaiming the ultimate future of the human species and the universe as a whole. As Dr. Atl puts it in *El futuro de la especie*: "The definitive future of men is not in reaching the perfection of any social or political system or in obtaining an organic, universal peace or in the integral reign of any religion. It is in their physical radiation onto other worlds" ("El futuro definitivo del hombre no está en alcanzar la perfección de cualquier sistema social o político ni en la obtención de una paz orgánica, universal ni en el reinado integral de alguna religión. Está en su radiación física sobre otros mundos").[139] Ultimately, conquering the universe "will transform Homo Sapiens into superman" ("transformará al *Homo Sapiens* en superhombre") and on this possibility "depends the future of the Earth" ("depende el porvenir de la Tierra").[140] During the final years of his life, Dr. Atl was deeply disillusioned about humanity's inability to accomplish this supreme goal due to what he considered scientific shortcomings and political obstacles. He died in 1964, convinced that the future of humanity as a biological species, geological agent, and cosmological entity relied on the construction of Olinka.

CONCLUSION

This chapter has examined Dr. Atl's and Nahui Olin's views on the inorganic vitality of the Earth and the cosmos. Both of them attributed lively behaviors—"conscious," deliberate, self-organizing, self-sustaining activities—to phenomena traditionally deemed "inert" such as volcanic activity, electromagnetic radiation, or outer space events. According to them, this vital nature of the cosmos was deeply intertwined with human beings by means of a direct, bodily interaction between cosmological forces and the human brain. Intelligence—one of the most powerful forces in the universe—should not be considered the result of a process of perception or representation, but the immediate "radiation" or exteriorization of cerebral matter. This unconventional understanding of the brain and its relationship with the universe shows how both Dr. Atl and Nahui Olin simultaneously build on and exceed scientific findings of the time regarding physics, cosmology, and geology. They strove to intervene in fundamental scientific debates, while also suggesting that established science was ultimately a limited approach that had to be surpassed by human intelligence. In the end, they aspired to create a "new" or "alternative" science that would not only generate an accurate understanding of the universe, untainted by inadequate scientific notions, but more importantly a future transformation of the cosmos.

Both Dr. Atl and Nahui Olin had high aspirations of producing a utopian line of flight that could originate a new universe. An exploration of the full potential of the brain was one of the main lines of speculation pursued by both of them. Whereas Nahui Olin envisioned releasing massive amounts of energy by splitting atoms in cerebral matter, Dr. Atl imagined that human brains could summon cosmic forces and produce materializations such as spacecrafts. Along with probing into the brain's potential, the prospect of space exploration entailed the most intriguing promise for both of them, because it offered the possibility of reaching unexplored places and physically altering them. For Nahui Olin, space exploration represented the greatest way of rebelling against the uniform, monotonous fabric of the universe in search of unprecedented realities. In a similar vein, Dr. Atl first privileged an understanding of space travel as a permanent, endless drive to explore the "beyond" of the universe and human knowledge; decades later, the project of Olinka reformulated his interest on space exploration as an anthropocentric need to conquer and control the cosmos—an aspiration that was already present in his novel—with clear fascist overtones. Overall, the imagination of Dr. Atl and Nahui Olin produced a comprehensive image of the interpenetration between human beings, geological processes, and cosmological phenomena, while also proposing how human intelligence should aspire to recreate the universe as a whole.

Epilogue

In 1971, after more than five years of work, the artist David Alfaro Siqueiros inaugurated his monumental mural cycle *La marcha de la humanidad en la tierra y hacia el cosmos* at the Siqueiros Cultural Polyforum in Mexico City. This impressive mural composed of four sections portrays a historical narrative marked by humanity's conflict both with social factors (such as slavery, oppression, and poverty) and natural components (volcanoes, plants, outer space) in search of a utopian future. The first section "La marcha de la humanidad hacia la revolución democrática burguesa" (south wall) depicts in subdued colors the relentless march of humanity represented by mothers and their children who face adverse factors such as the lynching of a Black enslaved person, the oppression of Indigenous peoples, a deceiving demagogue wearing a clown mask, and repressive and violent military actions. The second section "La marcha de la humanidad hacia la revolución del futuro" (north wall) follows the continuing procession of humanity in close interrelation and conflict with natural forces (erupting volcano, dead trees, barren land), which by the end of the section announces the regeneration of life suggested in bright colors by the flowering of an *amate* tree. The third section consists of two walls facing each other across the room—"El hombre" (west wall) and "La mujer" (east wall)—which according to Siqueiros represent the feminine (culture, harmony, peace) and masculine (science, technology, industrialization) principles guiding humanity's future. The final and unnamed section (ceiling) completes the mural's historical narrative by portraying abstract figures that represent astronauts wearing helmets and operating mechanisms, surrounded by circles resembling planets and other bodies in space.

FIGURE E.1. David Alfaro Siqueiros's *La marcha de la humanidad en la tierra y hacia el cosmos* (north wall and ceiling). Courtesy of Schalkwijk / Art Resource, NY. David Alfaro Siqueiros, © 2024 Artists Rights Society (ARS), New York / SOMAAP, Mexico City

At the time of the mural's inauguration, the septuagenarian Siqueiros was the last living muralist of what were called "Los tres grandes" after Rivera's and Orozco's death more than a decade earlier. Muralism itself—and even postrevolutionary culture as a whole—had lost the political and aesthetic avant-garde position that it had occupied in the 1920s and 1930s, becoming the bedrock of official culture during the PRI regime. At the same time, the 1969 moon landing by American astronauts had made space exploration a concrete possibility and a suitable objective for the future. Thus, Siqueiros's mural cycle embodies perhaps the last iteration of postrevolutionary biocosmism by adapting it to the age of space exploration and commercial spectacularization.[1] As its name suggests, this mural is not staged on a specific setting—a particular country or region such as Mexico or Latin America—but more on a planetary, cosmic setting.[2] Rather than conceiving the human being in relation to a national community and state sovereignty, the mural posits humanity as a biological, cosmic agent that enters into conflict with forces within human society and the natural world at large. In contrast with largely allegorical readings of the mural, I argue that geological and biological processes are not portrayed as a passive backdrop to human activity or as merely metaphorical figures, but as lively forces that materially challenge and interpenetrate human beings, forming "el drama de la vida."[3] As suggested by Siqueiros and a group of collaborators who created a sound and light show around the mural cycle:

> It is the book of life that begins or ends wherever it is opened . . . it is the fire that emerges from an explosion at the center of the earth, forming a volcano that floods the entire horizon with red and orange . . . and it is the newly green *amate*, flowering and exuberant, nourishing with hope the human race that is only at this one time at the very center of all the drama . . . the drama of life itself, which is a continuous birth.

> *Es el libro de la vida que empieza o termina donde quiera que se le abre . . . es el fuego que emerge con un estallido del centro de la tierra formando un volcán que inunda de rojo y naranja todo el horizonte . . . y es el amate reverdecido, florido y exuberante, nutriendo de esperanza al género humano que es, por esta única vez, el centro mismo de todo el drama . . . el drama de la vida misma, que es un parto continuo.*[4]

As with other biocosmists studied in this book, Siqueiros's utopian outlook displays a clear directionality by culminating with the depiction on the ceiling of what he calls "the conquest of the Cosmos to give rise to the man of the future, whose image, still amorphous, is the herald of a new and unknown model of life" ("la conquista del Cosmos para dar origen al hombre del futuro, cuya efigie, amorfa todavía, es heraldo de un nuevo y desconocido modelo de vida").[5] In the vein of Dr. Atl's Olinka project, space travel and exploration is construed as humanity's next evolutionary stage, but since humanity is considered a cosmic entity closely interrelated with geological and biological factors, space exploration indeed represents the future of the planet and of universal life itself. This is why the human conquest of outer space will bring about "a new and unknown model of life," that is, a new phase of cosmic vitality that will continue its journey of creation and transformation initiated since the beginning of the universe. In this process, not unlike the work of other Mexican biocosmists, Siqueiros's mural attributes a great deal of importance to science and technology as harbingers of cosmic progress, as represented in the astronauts' abstract, machinelike figures controlling mechanical devices: "Science and technology, leading to industrialization in creative hands that contribute their energy and knowledge to forge the new world" ("La ciencia y la técnica, que llevan a la industrialización en unas manos creadoras que entregan su energía y saber para forjar el mundo nuevo").[6]

In this way, Siqueiros's mural provides an apt illustration of the two interconnected arguments—one historical and the other conceptual or theoretical—that I have posited throughout this book. First, I have argued that biocosmism represents a minoritarian intellectual trend that was overshadowed by predominant motivations such as nation building in postrevolutionary Mexico. Biocosmic thought was embedded in the biopolitical reflection and political culture established after the Revolution, yet it simultaneously endeavored to contest their anthropocentric assumptions by focusing on the vitality of nonhuman beings and adopting a cosmic outlook. Siqueiros is a clear example of a postrevolutionary artist who, while mostly known for his experimental mural painting devoted to a Marxist-inspired critique of society and the advancement of proletarian ideals, produced art with evident biocosmic overtones that have been rarely examined. *La marcha de la humanidad* is perhaps his most overtly biocosmic work, yet one could also distinguish similar connotations in previous murals such as *Nueva democracia* (1944) and *Retrato de la burguesía* (1939–1940): while the former features the allegorical figure of a nude woman emerging from a volcano crater against a cosmic backdrop, the latter represents colossal electrical towers and other industrial structures rising to the heavens and reaching the sun pictured at the staircase ceiling of the Sindicato Mexicano de Electricistas. In this book, rather than aspiring to provide an exhaustive, definitive examination of biocosmism in postrevolutionary Mexico, my aim has been to examine a set of basic interests and ideas shared by a handful of intellectuals and artists so as to open an innovative space of inquiry that could be further explored.

In addition to this historical argument, I have argued that there is a constitutive theoretical tension that runs throughout biocosmic thought and is similarly replicated in Siqueiros's mural: on the one hand, this mural creates a historical narrative spearheaded by humans as the only beings capable of mastering the universe through the development of science and technology; at the same time, as I argued earlier, Siqueiros's work seems to suggest that humanity is nothing more than a concrete expression of cosmic vitality that, rather than aspiring to exert control over nature, strives to produce a radically new, utopian "model of life." In this book, I have tried to highlight this tension and analyze how it manifests itself in diverse ways in each of the intellectual projects of the Mexican biocosmists. As Frederic Jameson has suggested, the fact that utopian imagination remains inextricably linked to the ideological conditions of its formulation—apparently incapable of providing a true image of radical difference—opens up the possibility of understanding utopia as a defamiliarizing strategy.[7] The main function of utopian imagination as a conceptual and political mechanism would not reside in anticipating the future through meticulous programs, but in confronting our own ideological and epistemological limitations and thus defamiliarizing the categories that ground our comprehension of reality. Throughout the book I have showed that all biocosmic utopian *programs* remain entrenched in anthropocentric, biopolitical frameworks, often framing nature as an external object that humans can control and exploit, or reinforcing eugenic—even Nazi in the case of Vasconcelos and Dr. Atl—ideologies pervasive in Mexico and beyond. Paradoxically, the subtending utopian *impulse* of biocosmism was oriented toward defamiliarizing and destabilizing traditional ontological, biopolitical norms and divisions that organize our comprehension of the universe: what are the defining traits of human beings, what are the essential differences between humans and other biological beings, what is the fundamental distinction between the living/organic and the nonliving/inorganic.

By establishing a dialogue between biocosmism and recent new materialist, posthumanist approaches without collapsing one into the other, I have sought to underscore the continuing relevance of biocosmism in current debates both as a precursor and as a magnifying lens to examine the constitutive tensions of post-anthropocentric frameworks. Both biocosmism and the most recent theoretical trends, as iterations of the same intellectual tradition, endeavor to clear a pathway for a new understanding of the agency of matter, animals, plants, geological and cosmological phenomena, which in turn requires a reformulation of the *humanness* of humanity. But in this process they encounter the ongoing resistance of ingrained tensions within their own conceptualizations. In other words, while they strive to dismantle anthropocentric frameworks in search of a new configuration of the world, they often remain rooted in deep-seated ontological premises underlying the philosophical tradition of humanism. This is evident in the fact that biocosmists often employ humanist terms such as "ethics" or "consciousness" in order to theorize matter's powers of agency. As Jacques Lezra puts it in his sweeping critique of new materialism, "humanism returns for the New Materialism in 'nonhuman' form as matter."[8] Rather than deem it an inherent defect, this fundamental tension or contradiction reveals the sustained, utopian work required to move thinking beyond its hegemonic categories and premises in search of radically new ways of understanding and relating to the universe. Are humans truly *capable* of imagining a posthuman conceptualization and relationship with plants, animals, rocks, and the whole cosmos? This question, today as in the historical time of biocosmism, remains fundamentally unanswered and thus continues to trigger the inexhaustible drive of the utopian imagination.

Biocosmism has delved into the intense relationship between utopian thought and a broad theorization of vitality in postrevolutionary Mexico. During these transformative decades, the sustained problematization of life coupled with utopian desire produced speculative rearrangements of the vitality of the cosmos that to this day continue to challenge the limits of what we think is possible and impossible. The experimental work of Herrera, Vasconcelos, Nahui Olin, and Dr. Atl, among others, endeavored to suspend traditional conceptions and interrelations between humans, nonhuman forces and entities, and the cosmos as a whole, allowing for the possibility of a post-anthropocentric thought and a reformulation of biopower. In the end, biocosmism still resonates with us today and calls us to imagine the unpredictable ways in which we could inhabit otherwise the Earth and the universe.

Notes

INTRODUCTION

1. Diego Rivera, "El Anfiteatro de la Escuela Nacional Preparatoria," 82–84.
2. Rivera, "El Anfiteatro," 82–84.
3. Adrián Soto Villlafaña, "Diego Rivera: *La creación*," 18–23.
4. Dina Comisarenco Mirkin, "*La creación* by Diego Rivera," 36–37.
5. Leonard Folgarait, *Mural Painting and Social Revolution in Mexico, 1920–1940*, 40.
6. One important hub for cosmic thought was Soviet Russia, which had a cosmist tradition reaching back to the nineteenth-century thinker Nikolai Fedorov. In fact, the International Biocosmic Association came in contact with Russian cosmists such as Konstantin Tsiolkovsky and Alexander Chizhevsky, although it is unclear what kind of relationship, if any, it had with the short-lived Russian collective of Immortalist-Biocosmists led by Aleksandr Sviatogor. For an overview of Russian Cosmism in English, see George M. Young, *The Russian Cosmists: The Esoteric Futurism of Nikolai Fedorov and his Followers*. See also the anthology edited by Boris Groys, *Russian Cosmism*.
7. Horacio Legrás, *Culture and Revolution*, 1–2.
8. For an overview of the period, see Horacio Legrás, *Culture and Revolution*; and Jorge Aguilar Mora, *Una muerte sencilla, justa, eterna*; as well as *The Eagle and the Virgin: Nation and Cultural Revolution in Mexico, 1920–1940*, edited by Mary Kay Vaughan and Stephen E. Lewis. For a transnational perspective, see Adela Pineda Franco, *The Mexican Revolution on the World Stage*.
9. Carolyn Fornoff, "Planetary Poetics of Extinction in Contemporary Mexican Poetry," 231.
10. Fornoff, "Planetary Poetics of Extinction," 231.
11. Cristina Rivera Garza, *Escrituras geológicas*, 10–14.
12. See Rivera Garza, *Escrituras geológicas*.

13. Rivera Garza, *Autobiografía del algodón*, 83–92; *Escrituras geológicas*, 35–53. In addition to Rivera Garza's and Fornoff's work, Mary Louise Pratt's recent book also explores Latin American culture from a planetary perspective akin to biocosmism. See Pratt, *Planetary Longings*. See also Emily Celeste Vázquez Enríquez, "Planetariedad en 'El mal de la taiga' de Cristina Rivera Garza."
14. Eli de Gortari, *La ciencia en la historia de México*, 324, 374; Elías Trabulse, *Historia de la ciencia en México: Estudio y textos. Siglo XIX*, 13; Ruy Pérez Tamayo, *Historia general de la ciencia en México en el siglo XX*, 57.
15. Enrique Beltrán, *Medio siglo de ciencia mexicana, 1900–1950*, 34–36. Ismael Ledesma-Mateos, *De Balderas a la Casa del Lago: La institucionalización de la biología en México*, 28–29.
16. Roberto Moreno, *La polémica del darwinismo en México*, 37–38; Rosaura Ruiz Gutiérrez, *Positivismo y evolución: Introducción del darwinismo en México*, 83. Bill Mesler and H. James Cleaves, *A Brief History of Creation: Science and the Search of the Origin of Life*, 148.
17. One of the rare exceptions is Ruiz Gutiérrez, *Positivismo y evolución*, 83–137. Ruiz Gutiérrez characterizes Herrera as "one of the few biologists [in Mexico] with a clear and conscious philosophical conception" ("de los pocos biólogos [en México] con una concepción filosófica clara y consciente") (83).
18. The bibliography on the Ateneo de la Juventud is extensive and diverse. For an up-to-date introduction and analysis of the standard interpretations of the Ateneo's significance in Mexican culture, see Alfonso García Morales, "El Ateneo de México: Crónica e interpretación de un proyecto intelectual," 127–55. See also Pedro Ángel Palou, "The Ateneo de la Juventud: The Foundations of Mexican Intellectual Culture," 233, 246.
19. This line of argumentation was perhaps inaugurated by Samuel Ramos, a founding figure of the "filosofía de lo mexicano" with his *Perfil del hombre y la cultura en México* (1934). See Samuel Ramos, *Historia de la filosofía en México*, 137–48. Other instances of this interpretation can be found in influential texts such as Luis Villoro, "La cultura mexicana de 1910 a 1960," *Historia Mexicana* 10. 2 (1960), 198–200; José Gaos, "En torno a la filosofía mexicana" in *Obras completas VIII*, 303, 322. Abelardo Villegas, *La filosofía de lo mexicano*, 97–99. A recent English language anthology of Mexican philosophy supports this reading as well: *Mexican Philosophy in the 20th Century: Essential Readings*, edited by Carlos Alberto Sánchez and Robert Eli Sanchez Jr, xxvii.
20. Cuauhtémoc Medina, *Olinka: La ciudad ideal del Dr. Atl*, 29–31. Carmen Gaitán Rojo, "Los apreciables jóvenes del Ateneo de la Juventud" in *El Ateneo de la Juventud y la plástica mexicana*, 19–23.
21. See, for example, Antonio Luna Arroyo, *El Dr. Atl: Sinopsis de su vida y su pintura*, 156–61. More recently, see Amaya Larucea Garritz, *País y paisaje: Dos invenciones del siglo XIX mexicano*, 181–82.

22. For an excellent overview of postrevolutionary literature, see *La revolución intelectual de la Revolución mexicana (1900–1940)*, edited by Yanna Hadatty Mora, Norma Lojero Vega, and Rafael Mondragón Velázquez.
23. María del Carmen Rovira Gaspar, *Una aproximación a la historia de las ideas filosóficas en México: Siglo XIX y principios del XX. Tomo I*, 482–84. Among the thinkers included in this nineteenth-century philosophical trend, Juan Nepomuceno Adorno has the greatest affinity with biocosmic thought, including its utopian component and central notion of universal life. See Pablo González Casanova, *Un utopista mexicano*; María del Carmen Rovira Gaspar, *Dos utopistas mexicanos del siglo XIX: Francisco Severo Maldonado y Ocampo y Juan Nepomuceno Adorno*.
24. Rubén Gallo, *Mexican Modernity: The Avant-Garde and the Technological Revolution*, 10.
25. Horacio Legrás, *Culture and Revolution: Violence, Memory and the Making of Modern Mexico*, 3.
26. Joshua Lund, *The Mestizo State: Reading Race in Modern Mexico*, XV.
27. Pedro Ángel Palou, *El fracaso del mestizo*, 14–15. See also Rebecca Janzen, *The National Body in Mexican Literature: Collective Challenges to Biopolitical Control*.
28. David Dalton, *Mestizo Modernity*, 4–5. Susan Antebi, *Embodied Archive*, 2–3.
29. Mary Coffey, *How a Revolutionary Art Became Official Culture*, 11–12; Adriana Zavala, *Becoming Modern, Becoming Tradition: Women, Gender, and Representations in Mexican Art*.
30. Michel Foucault, "Right of Death and Power over Life." *Biopolitics: A Reader*, 46. This reader edited by Campbell provides an excellent overview of the main discussions and themes in biopolitical reflection. See also *Excesos de vida: Ensayos sobre biopolítica*, edited by Garbiel Giorgi and Fermín Rodríguez.
31. Foucault, "Right of Death and Power over Life," 43.
32. Foucault, "Right of Death," 47.
33. Foucault, *Society Must Be Defended: Lectures at the Collège de France, 1975–1976*, 254, 255.
34. For a Latin American perspective, see Daniel Nemser, "Biopolitics in Latin America," in *The Encyclopedia of Postcolonial Studies*, 178–84. As Nemser argues, "*Mestizaje* may well be the most distinctive form that biopolitics has taken in Latin America" (180). See also *Heridas abiertas: Biopolítica y representación en América Latina*, edited by Mabel Moraña and Ignacio Sánchez Prado.
35. In 1911, 303 Chinese people were murdered in the town of Torreón at the hands of revolutionary troops. Later, in the 1920s and 1930s anti-Chinese campaigns and laws appeared in Sonora and other states advancing the idea of Chinese people as a biological, economic, and social threat. See Jorge Quintana Navarrete, "Sinofobia y mestizaje eugenésico en la prensa de Sonora."
36. Giorgio Agamben, *Homo Sacer*, 8–11.

37. For an overview of eugenics in Mexico see Susan Antebi, *Embodied Archive*; Nancy Stepan, *The Hour of Eugenics: Race, Gender and Nation in Latin America*; Laura Suárez y López Guazo, *Eugenesia y racismo en México*; and Beatriz Urías Horcasitas, *Historias secretas del racismo en México (1920–1959)*.
38. Foucault, *Society Must Be Defended*, 255.
39. The contradiction between biopolitical and posthuman impulses in biocosmism echoes the tension that Susan Antebi explores in *Embodied Archive*, which analyzes the same period through a disability studies perspective: the tension between the biopolitical control and optimization of human corporeal differences, and the contingent, material affirmation of those same differences, which were always at risk of being subsumed into state or medical frameworks.
40. These theoretical trends are burgeoning fields of reflection in US academia and—to a lesser degree—in Latin American countries. For an introduction to these fields see: *New Materialisms: Ontology, Agency, and Politics*, edited by Diana Coole and Samantha Frost; *Heridas abiertas*, edited by Mabel Moraña and Ignacio Sánchez Prado; Matthew Calarco, *Thinking through Animals: Identity, Difference, Indistinction*; Gabriel Giorgi, *Formas communes: Animalidad, cultura, biopolítica*; Jeffrey T. Nealon, *Plant Theory: Biopower and Vegetable Life*.
41. See chapter 1 of Nealon, *Plant Theory: Biopower and Vegetable Life*. See also Cary Wolfe, *Before the Law: Humans and Animals in a Biopolitical Frame*, which takes on the task of "radicaliz[ing] biopolitical thought in ways no possible if we remain within the usual purview of anthropocentrism" (52). For a reflection on the "anthropological machine" producing the idea of "the human," see Giorgio Agamben, *The Open: Man and Animal*, 13–16. For an analysis on how biopower relies on the distinction between life and nonlife, see Elizabeth Povinelli, *Geontologies: A Requiem to Late Liberalism*. As Povinelli contends, "biopower (the governance through life and death) has long depended on a subtending geontopower (the difference between the lively and the inert)" (5).
42. Christopher N. Gamble, Joshua S. Hanan and Thomas Nail. "What Is New Materialism?," 120. This article does a great job of summing up the main critiques levelled against new materialist theories.
43. Rosi Braidotti, *Posthuman Knowledge*, 52–61.
44. Alan Knight, "El utopismo y la Revolución Mexicana," 88, 115.
45. Knight, "El utopismo," 88.
46. Sargent Lyman Tower, "Three Faces of Utopianism Revisited," 4; Fredric Jameson, *Archaeologies of the Future: The Desire Called Utopia and Other Science Fictions*, 2–4; Ruth Levitas, *The Concept of Utopia*, 209. See also Fátima Vieira, "The Concept of Utopia," 6–7.
47. Lyman Tower, "Three Faces of Utopianism Revisited," 4.
48. In addition to Dr. Atl's Olinka, teacher José de la Luz Mena founded in 1917 a school/colony in Chuminópolis, on the outskirts of Mérida, Yucatán, in which he tried to practice the educational ideals of the Rationalist School.

See Belinda Arteaga Castillo, *La escuela racionalista en Yucatán: Una experiencia mexicana de educación anarquista (1915–1923)*. Juana Belén Gutiérrez de Mendoza—an outstanding journalist and anarchist, feminist activist—established in 1919 the Colonia Agrícola Experimental Santiago Orozco in the state of Morelos, a community named in honor of her late Zapatista friend. See Alicia Villaneda, *Justicia y libertad: Juana Belén Gutierrez de Mendoza (1875–1942)*. English writer D. H. Lawrence traveled to Mexico between 1923 and 1925 in search of a place for founding a commune removed from the decline of Western capitalism. See Armando Pereira, "D. H. Lawrence: México, la utopía imposible." In 1926, the Estridentista vanguard—particularly Germán List Arzubide—envisioned the transformation of Xalapa, Veracruz, into Estridentópolis, a hypermodern and proletarian city. See Lynda Klich, "Estridentópolis: Achieving a Post-revolutionary Utopia in Jalapa." In 1938, the Unión de Arquitectos Socialistas presented an architectural project to build a utopian city based on cooperative principles. See Alfonso Fierro, "Modeling the Urban Commune: Collective Housing, Utopian Architecture, and Social Reproduction in the Mexican 1930s." In addition to all these cases, there were several other projects of founding utopian communities inspired by the Paris Commune in postrevolutionary Mexico. See Bruno Bosteels, *La comuna mexicana*.

49. Regarding Urzaiz's novel, see chapter 1 of Susan Antebi's *Embodied Archive* and Jorge Quintana Navarrete, "La utopía a prueba: Formas heterogéneas de vida en 'Eugenia' de Eduardo Urzaiz." In addition to Urzaiz's novel, Ricardo Flores Magón published anarchist utopias in his short story "El sueño de Pedro" (1912) and in a 1922 letter to Ellen White, in which he described "La Ciudad de la Paz." See chapter 2 of Benjamín Maldonado, *La utopía magonista*. Carrancista general and then governor of Yucatán Salvador Alvarado in 1916 published *Mi sueño*, a utopia guided by early socialist principles. See Jorge Quintana Navarrete, "El sueño de Salvador Alvarado: Socialismo utópico y subalternidad en el Yucatán Revolucionario." Djed Bórquez published the novel *Yorem Tamegua* (1923), in which he imagines the foundation of an agricultural cooperative by Yaqui and Mayo peoples in Sonora in 1935. See Karel Van Horn Kopka, "La literatura sonorense en la década de los veinte: Caso concreto: Yorem Tamegua." The dystopian subgenre also appeared in postrevolutionary Mexico: José Vasconcelos published the anti-Semitic, anti-Communist short story "México en 1950" in his work *Qué es la revolución* (1937), while Félix Palavicini published his novel *¡Castigo! Novela mexicana de 1945* (1926), in which a brutal communist coup transforms Mexico into the República Soviet de Obreros, Soldados y Campesinos.

50. In addition to Vasconcelos's *La raza cósmica*, Constitutionalist Tomás Rosales published the utopian project *El gobierno de mañana: República social sinárquica* (1915), in which he outlines a harmonious government at the height of the civil war between revolutionary factions. Zapatist Miguel

Mendoza López Schwertfeger devised throughout his life several socialist plans for radical transformation under different names: República Social de Trabajo, Gran Federación de Municipios de la República Mexicana, República Industrial Socialista, Liga de las Clases Productoras. See Patricia Valles, *Del anarquismo a la utopía: La visión revolucionaria de Miguel Mendoza Lopez Schwertfeger*. Additionally, Alfonso Reyes endeavored to create a history and reflection of Latin America utopianism in works such as *No hay tal lugar* and *Última tule*. See Ignacio Sánchez Prado, "The Age of Utopia: Alfonso Reyes, Deep Time and the Critique of Colonial Modernity."

51. Pedro Ángel Palou, *El fracaso del mestizo*, 14.
52. I take the distinction between utopian program and impulse from Fredric Jameson, *Archaeologies of the Future: The Desire Called Utopia and Other Science Fictions*, 2–4.
53. Miguel Abensour, "Persistent Utopia," 407.

CHAPTER 1

1. Ismael Ledesma-Mateos, *De Balderas a la Casa del Lago: La institucionalización de la biología en México*, 28–29.
2. Henderson James Cleaves et al, "Introduction," 3.
3. Bill Mesler and Henderson James Cleaves, *A Brief History of Creation: Science and the Search of the Origin of Life*, 148; Ruiz Gutiérrez, *Positivismo y evolución*, 99.
4. Cleaves et. al, "Introduction," 1–2; Alfonso L. Herrera, *Nociones de biología*, 14.
5. Antonio Lazcano, "What Is Life?: A Brief Historical Overview," 5–6.
6. Mauricio de Carvalho Ramos, *A plasmogenia e a síntese conceitual e artificial do protoplasma*, 15–16.
7. de Carvalho Ramos, *A plasmogenia*, 36.
8. Herrera, *Nociones de biología*, 134.
9. Ruiz Gutiérrez, *Positivismo y evolución*, 97–98.
10. Pastor Rouaix, "La Dirección de Estudios Biológicos y la obra del Profesor Alfonso L. Herrera," 232.
11. Palou, *El fracaso del mestizo*, 15.
12. Gareth Williams, *The Mexican Exception*, 4–5.
13. Vaughan and Lewis, introduction to *The Eagle and the Virgin*, 10.
14. Manuel Gamio, *Forjando patria*, 26.
15. See Agustín Francisco Basave Benítez, *México mestizo: Análisis del nacionalismo mexicano en torno a la mestizofilia de Andrés Molina Enríquez*. See also Joshua Lund, *The Mestizo State: Reading Race in Modern Mexico*.
16. Herrera, "La biología en México durante un siglo," 7–8.
17. Herrera, "La biología en México durante un siglo," 16; Herrera, "Inauguración de la Dirección de Estudios Biológicos," 10.
18. Herrera, *La Plasmogenia: Nueva ciencia del origen de la vida*, 15.
19. Herrera, *Una nueva ciencia: La plasmogenia*, 82.

20. Herrera, *Una nueva ciencia: La plasmogenia*, 81–82.
21. Herrera, "Estudios de plasmogenia," 54.
22. Herrera, "Estudios de plasmogenia," 55.
23. Herrera, "Los cerebros artificiales," 314.
24. Castellanos, Israel. *La plasmogenia*, 41.
25. Herrera, *Una ciencia nueva, la plasmogenia*, 82.
26. Israel Castellanos, *La plasmogenia*, 124.
27. Herrera, "La aspiración gigantesca de la plasmogenia," 18.
28. Herrera, *Biología y plasmogenia*, 309.
29. Castellanos, *La plasmogenia*, 124.
30. Agamben, *Homo Sacer*, 6.
31. Peter Sloterdijk, "Reglas para el Parque Humano," 15.
32. Castellanos, *La plasmogenia*, 123.
33. Roberto Esposito, *Bíos: Biopolitics and Philosophy*, 127.
34. Herrera, "Inauguración de la Dirección de Estudios Biológicos,"13.
35. Herrera, "Les musées de l'avenir," 226.
36. Herrera, *Guía para visitar el Museo Nacional de Historia Natural*, 4.
37. Herrera, *Guía para visitar*, 4.
38. Ruiz Gutiérrez, *Positivismo y evolución: Introducción del darwinismo en México*, 87.
39. Herrera, *Una nueva ciencia: La plasmogenia*, 28.
40. Herrera, *Una nueva ciencia*, 15.
41. Herrera, *Una nueva ciencia*, 28.
42. Herrera, *Una nueva ciencia*, 406.
43. Herrera, *Una nueva ciencia*, 287.
44. Herrera, *Una nueva ciencia*, 287.
45. Herrera, *Nociones de biología*, 216; Ruiz Gutiérrez, *Positivismo y evolución*, 127–28.
46. Herrera, *Nociones de biología*, 243.
47. Herrera, *Una nueva ciencia*, 398; Ruiz Gutiérrez, *Positivismo y evolución*, 88.
48. Herrera, *La filosofía etérea*, 17.
49. Herrera, *La Plasmogenia: Nueva ciencia del origen de la vida*, 30–32. Emphasis in original.
50. Herrera, *Una nueva ciencia*, 195.
51. Herrera, *Una nueva ciencia*, 266.
52. See, for example, *Précis de Solidarité Biocosmique* (1928) by Albert Mary and Félix Monier.
53. Herrera, *Una nueva ciencia*, 294.
54. The notion of biocosmic solidarity might be considered an attempt to extend anarchist solidarity to the whole cosmos. In fact, the idea of universal life and plasmogeny circulated predominantly in anarchist publications in Argentina, France, and Spain. Herrera's work *Filosofía etérea*, published in Argentina in 1919, also includes an article by the Argentine anarchist Severo Bruno who

argues that plasmogeny "enriches the experimental conclusions of the scientific encyclopedia of anarchy" ("la Plasmogenia, y por ende su hija, la filosofía Etérea, enriquecen con sus conclusiones experimentales la enciclopedia científica de la Anarquía"). Bruno acknowledges, however, that plasmogeny "does not solve the social-political problem of peoples, a fundamental cause of their unhappiness" ("no resuelve el problema social-político de los pueblos, causa fundamental de la infelicidad de los mismos"). For its part, the anarchist journal *Estudios*, published in Valencia, Spain, between 1928 and 1937, published several articles on plasmogeny to reinforce a materialist conception of the world. Likewise, in the *Enclopédie anarchiste* (1934), which surveys the main theories and figures that have broadened the intellectual tradition of anarchism, the author Sébastien Faure included two separate articles on plasmogeny and biocosmic solidarity.

55. Gilles Deleuze and Félix Guattari, *A Thousand Plateaus: Capitalism and Schizophrenia*, 499.
56. Gilles Deleuze, "La inmanencia" 40.
57. See *El derecho a volar: Alfonso Reyes y la ciencia*, edited by Margarito Cuéllar.
58. Robert T. Cohn, "Official Nationalism in Mexico: Alfonso Reyes and the Hispanization of High Culture at the Turn of the Century," 105–6.
59. Alfonso Reyes, *Obras completas: Tomo VIII*, 409.
60. Reyes, *Obras completas: Tomo VIII*, 409.
61. Reyes, *Obras completas: Tomo VIII*, 410.
62. Reyes, *Obras completas: Tomo VIII*, 409.
63. Reyes, *Obras completas: Tomo VIII*, 409.
64. Sheldon Penn, "*Visión de Anáhuac* (1519) as Virtual Image: Alfonso Reyes's Bergsonian Aesthetic of Creative Evolution," 131.
65. Cohn, "Official Nationalism in Mexico,"106.
66. Reyes, *Obras completas: Tomo VIII*, 411.
67. Reyes, *Obras completas: Tomo VIII*, 411.
68. Reyes, *Obras completas: Tomo VIII*, 410.
69. Ignacio Sánchez Prado, "Las reencarnaciones del centauro: *El deslinde* después de los estudios culturales," 27–30.
70. Gareth Williams, *The Mexican Exception: Sovereignty, Police, and Democracy*, 112.
71. Vicente Huidobro, "La creación pura," 111.
72. Deleuze, "Literature and Life," 1.
73. Elizabeth Grosz, *Chaos, Territory, Art: Deleuze and the Framing of the Earth*, 1–24.
74. Deleuze and Guattari. *What Is Philosophy*, 24.
75. In fact, Mexican bioartists Tania Aedo and Federico Hemmer founded the Membrana Lab in Mexico City, a facility for experimenting with biotechnology, bioart, and extended reality. In this facility, they reproduced and exhibited Herrera's plasmogenic experiments. See "What Is the Purpose of BioArt in a

VUCA World?: Introducing FabCafe Mexico City with Georg Tremmel (Artist), Ana Laura Cantera (Bioelectronic Artist)," FabCafe, February 16, 2022, https://fabcafe.com/events/20220216_bio_art_mexicocity
76. Robert Mitchell, *Bioart and the Vitality of Media*, 36–40.
77. Mitchell, *Bioart and the Vitality of Media*, 70.
78. Herrera, *Murmullos del universo*, 2.
79. Herrera, *Murmullos del universo*, 2.
80. Herrera, *Murmullos del universo*, 51.
81. Herrera, *Murmullos del universo*, 51.
82. Herrera, *Murmullos del universo*, 3.
83. There are several poems that allude to this notion in their titles: "Las voces del cosmos," "El coro eterno," and "Vagos rumores en la pradera," among others.
84. Herrera, *Murmullos del universo*, 107.
85. Herrera, *Murmullos del universo*, 103.
86. Herrera, *Murmullos del universo*, 103.
87. Herrera, *Murmullos del universo*, 2.
88. Herrera, *Murmullos del universo*, 25.
89. Herrera, *Murmullos del universo*, 37.
90. Herrera, *Murmullos del universo*, 38.
91. Herrera, *Murmullos del universo*, 120, 36.
92. Herrera, *Murmullos del universo*, 28.
93. Herrera, *Murmullos del universo*, 28.
94. Herrera, *Murmullos del universo*, 42.
95. Herrera, *Murmullos del universo*, 43.
96. Peter Bürger, *Theory of Avant-Garde*, 20–25.
97. Gallo, *Mexican Modernity*, 24.

CHAPTER 2

1. Ilja Nieuwland, *American Dinosaur Abroad: A Cultural History of Carnegie's Plaster Diplodocus*, 6.
2. Herrera, "Cómo obtuvo México el modelo del Diplodocus: Fue el antiguo Director de Estudios Biológicos quien lo solicitó," *El Universal*, April 11, 1930.
3. The Diplodocus was kept on display from 1931 to 1963 at the National Museum of Natural History. In 1964, the fossil was transferred to the Museum of Natural History, recently founded in the Chapultepec forest, where it is exhibited to this day. See Eduardo Vázquez Martín, "El viaje del Diplodocus," in *El Chopo año por año, 1975–2010*, 130.
4. Herrera, *El híbrido del hombre y el mono*, 8.
5. Herrera, *El híbrido del hombre*, 8.
6. Herrera, *El híbrido del hombre*, 13.
7. To learn more about Ivanov's experiments, see Kirill Rossianov's article which documents Ivanov's story using information from Soviet archives. Kirill Rossianov, "Beyond Species: Il'ya Ivanov and His Experiments on

Cross-Breeding Humans with Anthropoids Apes." See also Alexander Etkind, "Beyond Eugenics: the Forgotten Scandal of Hybridizing Humans and Apes."
8. Herrera, "Mi labor revolucionaria en la enseñanza," 57.
9. Herrera, "Mi labor revolucionaria," 57.
10. Herrera, *El híbrido del hombre*, 18.
11. Etkind, "Beyond Eugenics," 205.
12. Rossianov, "Beyond Species," 286.
13. Herrera, *El híbrido del hombre*, 5.
14. Herrera, *El híbrido del hombre*, 31.
15. According to Bantjes, the successive waves of revolutionary anticlericalism had three peak moments: the iconoclast attacks performed by the Constitutionalist military in 1914–1915, the "defanatizing" campaigns launched by regional leaders (Tomás Garrido Canabal, Adalberto Tejeda) in the 1920s, and the secularizing efforts of federal and provincial officials during the Maximato (1931–1935). Adrian A. Bantjes, "The Regional Dynamics of Anticlericalism and Defanatization in Revolutionary Mexico," in *Faith and Impiety in Revolutionary Mexico*, 111–21.
16. Ben Fallaw, "Varieties of Mexican Revolutionary Anticlericalism: Radicalism, Iconoclasm, and Otherwise, 1914–1935," 482.
17. Etkind, "Beyond Eugenics," 207.
18. Quoted in Herrera, *El híbrido del hombre*, 28.
19. Peter Wade, *Race and Ethnicity in Latin America*, 9.
20. Herrera, *El híbrido del hombre*, 24.
21. Herrera, *El híbrido del hombre*, 24.
22. Herrera, *El híbrido del hombre*, 26.
23. Herrera, *El híbrido del hombre*, 3.
24. Herrera, *El híbrido del hombre*, 4.
25. Calarco, *Thinking through Animals*, 5. Calarco maps the field of animal studies employing three concepts: identity, difference, and indistinction. The *identity* approach underscores the animal-human identity and tends to include a reflection on animal rights and animal liberation. The *difference* framework ponders the radical differences between humans and animals, and even among nonhuman animal species themselves, with the aim of destabilizing hegemonic conceptions of human nature. Lastly, the *indistinction* framework sheds light on the diverse relationships between specific animals and humans, emphasizing the shared traits of creativity and agency that unify them.
26. Leonard Finkelman, "De-extinction and the Conception of Species," 32.
27. The Tauros Programme is an initiative led by Dutch organizations and research centers that aims to backbreed the auroch (https://www.taurosproject.com). For its part, the Quagga Project was launched by Reinhold Rau in South Africa (https://quaggaproject.org/the-project).

28. In 2009, a Spanish team of researchers achieved the deextinction of the Pyrenean ibex using genomic transfer. An individual was born but died within a few minutes due to a lung defect.
29. In a 2013 interview, George Church, Harvard geneticist and leader of the so-called Lazarus Project aiming to resuscitate the wooly mammoth, claimed that the Neanderthal de-extinction will be achieved during his lifetime and that it needed "an extremely adventurous human female" to serve as a surrogate mother.
30. Claudio Campagna, Daniel Guevara, and Bernard Le Boeuf, "De-scenting Extinction: The Promise of De-extinction May Hasten Continuing Extinctions," S50.
31. Sophia Roosth, *Synthetic: How Life Got Made*, 170.
32. Shlomo Cohen, "The Ethics of De-Extinction," 165.
33. Herrera, *Una nueva ciencia: La plasmogenia*, 414. Herrera's ideas on this respect are akin to Russian cosmist Fedorov's project of technological resurrection of the dead. Certainly, Herrera was aware of the speculations of Russian cosmists, since some of its advocates participated in the journal of the biocosmic association he had founded in France. However, Herrera does not draw explicitly from the work of Fedorov. For an overview of Russian cosmists ideas, see the anthology edited by Boris Groys.
34. Herrera, *Una nueva ciencia*, 415.
35. Ruiz Gutiérrez, *Positivismo y evolución*, 91.
36. Herrera, *Una nueva ciencia*, 415.
37. Herrera, *Una nueva ciencia*, 415.
38. Garland E. Allen, "Mechanism, Vitalism and Organicism in Late Nineteenth and Twentieth-Century Biology," 261–83.
39. Herrera, "La filosofía etérea," 2.
40. Herrera, "La filosofía etérea," 2.
41. Herrera, "La filosofía etérea," 2.
42. Herrera, "La filosofía etérea," 2. The theory of ether is currently considered discredited by the scientific community. It is conventionally accepted that Einstein's theory of relativity (1915) fundamentally rejected the ether hypothesis that had predominated since Ancient Greece. However, it has been suggested that certain characteristics of the ether were preserved in the later concepts of "absolute space" (Newton), "electric field" (Maxwell), and "space-time" (Einstein). In this view, all those notions rest upon the assumption of a continuous medium that is indispensable for the interaction of bodies and matter; see Elaine Paiva de Andrade, Jean Faber, and Luiz Pinguelli Rosa, "A Spontaneous Physics Philosophy on the Concept of Ether throughout the History of Science: Birth, Death and Revival." Herrera continued to defend the existence of the ether into the 1920s and 1930s, although he sometimes substituted the ether for another basic element such as the electron.

43. Herrera, "La filosofía etérea," 2.
44. Herrera, La filosofía etérea, 20.
45. Herrera, La filosofía, 21.
46. Herrera, La filosofía, 21.
47. Herrera, La filosofía, 22.
48. Herrera, La filosofía, 37.
49. Herrera, La filosofía, 6, 3.
50. Herrera, Nociones de biología, 87.
51. Herrera, Una nueva ciencia, 383.
52. Herrera, "El origen del pensamiento," 24.
53. Van Drunen quoted in Herrera, Una nueva ciencia, 383.
54. Van Drunen quoted in Herrera, Una nueva ciencia, 384.
55. Deleuze and Guattari, A Thousand Plateaus, 411.
56. Deleuze and Guattari, A Thousand Plateaus, 410.
57. Janet Bennet, Vibrant Matter: A Political Ecology of Things, 52–61
58. Herrera, Una nueva ciencia, 383.
59. Deleuze and Guattari, A Thousand Plateaus, 410–11.
60. On aleatory materialism, see Louis Althusser, Philosophy of the Encounter: Later Writings, 1978–87, 192.
61. Herrera, Una nueva ciencia, 423.
62. Elizabeth Povinelli, Geontologies: A Requiem to Late Liberalism, 45.
63. María del Carmen Rovira Gaspar, "El Ateneo de la Juventud," in Una aproximación a la historia de las ideas filosóficas en México, 883.
64. Guillermo Hurtado, La revolución creadora: Antonio Caso y José Vasconcelos en la Revolución Mexicana, 146.
65. Rosa Krauze de Kolteniuc, La filosofía de Antonio Caso, 80–81.
66. Antonio Caso, La existencia como economía, como desinterés y como caridad, 26.
67. Caso, La existencia, 25.
68. Caso, La existencia, 25.
69. Caso, La existencia, 23.
70. Caso, La existencia, 24.
71. Caso, La existencia, 26, 28.
72. Caso, La existencia, 37.
73. Caso, La existencia, 24.
74. Caso, La existencia, 35; Hurtado, La revolución creadora, 146–47.
75. Krauze de Kolteniuc, La filosofía, 96.
76. Caso, La existencia, 29.
77. Ledesma-Mateos, De Balderas a la Casa del Lago: La institucionalización de la biología en México, 29.
78. Ledesma-Mateos, De Balderas, 276.
79. Fernando Ocaranza, La tragedia de un rector, 89.
80. Ocaranza, La tragedia de un rector, 90.

81. Ocaranza, *La tragedia de un rector*, 93.
82. Ocaranza, *La tragedia de un rector*, 93.
83. Ocaranza, *La tragedia de un rector*, 99.
84. Ocaranza, *La tragedia de un rector*, 117.
85. Ocaranza, *La tragedia de un rector*, 122, 118.
86. Ocaranza, *La tragedia de un rector*, 114.
87. Ocaranza, *La tragedia de un rector*, 114.
88. Eliseo Ramírez Ulloa, "La simulación en la investigación biológica," 214.
89. Ramírez Ulloa, "La simulación," 214.
90. Ramírez Ulloa, "La simulación," 217.
91. Ramírez Ulloa, "La simulación," 214–15.
92. Ocaranza, *La tragedia de un rector*, 89.
93. Ramírez Ulloa, "La simulación," 103; Ocaranza, *La tragedia de un rector*, 104.
94. Ocaranza, *La tragedia de un rector*, 92.
95. Ocaranza, *La tragedia de un rector*, 99.
96. Emeterio Valverde Téllez, *Bibliografía filosófica: Tomo Segundo*, 135. Valverde Téllez was a Catholic priest and Bishop of León who wrote several works on the history of Mexican philosophy from a Christian point of view. In his *Bibliografía filosófica mexicana* (1913), Valverde Téllez dedicated a section to reviewing Herrera's institutional, scientific, and philosophical work, recognizing that he is a prolific scientist and an effective state official. However, according to Valverde Télez, it is unfortunate that "his philosophical ideas are materialistic and evolutionary" ("sus ideas filosóficas sean materialistas y evolucionistas") and full of "evident, solemn manifestations of irreligion" ("paladinas, solemnes manifestaciones de irreligión") (135).
97. Cleaves et al., "Introduction," 3. See also Enrique Beltrán, *Contribución de México a la biología: Pasado, presente y futuro*, 99–101.
98. From 1932 to 1942, Herrera published the *Bulletin du Laboratoire de Plasmogenie*, a monthly publication to share the products of his experiments.
99. See, for example, Adrián Soto Villafaña, "Diego Rivera: El hombre en el cruce de caminos." See also Coffey, *How a Revolutionary Art Became Official Culture*, 35–38.
100. Catha Paquette, *At the Crossroads: Diego Rivera and His Patrons at MoMA, Rockefeller Center, and the Palace of the Fine Arts*, 231–48.
101. Paquette, *At the Crossroads*, 222.
102. Coffey, *How a Revolutionary Art*, 46.
103. For an analysis of the role played by energy in this mural, see Renato González Mello, *La máquina de pintar*, 202.
104. Gallo, *Mexican Modernity*, 6. For discussion on the role of technology in conjunction with indigenismo in Rivera's work, see David Dalton, *Mestizo Modernity*, 81–98.
105. Adrián Villagómez, "La biología en el muralismo de Diego Rivera," 26.
106. Villagómez, "La biología," 26.

107. Paquette, *At the Crossroads*, 235.
108. Rivera's biographer Bertram Wolfe maintained that his political ideas at the time of his departure from Europe were "an undigested mixture of Spanish anarchism, Russian terrorism, Soviet Marxism-Leninism, Mexican agrarianism—the redemption of the poor peasant and the Indian." Wolfe, *The Fabulous Life of Diego Rivera*, 577.
109. For an account of Rivera's engagement in artistic circles during his stay in the Soviet Union, see Richardson, "The Dilemmas of a Communist Artist: Diego Rivera in Moscow, 1927–1928," as well as Mariano Meza Marroquín, "Diego Rivera y la experiencia en la URSS."
110. Meza Marroquín, "Diego Rivera y la experiencia en la URSS," 25.
111. Alexander Svyatogor, "Our Affirmations," 61.
112. Meza Marroquín, "Diego Rivera y la experiencia en la URSS," 19.
113. According to Lazcano, Rivera personally met Oparin in Moscow and made a drawing of him, but I have not found this portrait. Lazcano Araujo, "El agua y el origen de la vida," 93.
114. J. L. Bada and A. Lazcano, "Prebiotic Soup—Revisiting the Miller Experiment," 745–46.
115. According to Lazcano, in the 1930s Herrera read the English translation of Oparin's work and started an extended exchange of letters, which are now unfortunately lost. Lazcano Araujo, "El agua y el origen de la vida," 90.
116. Kathryn E. O'Rourke, "Gardens and Landscapes of Frida Kahlo's Mexico City," 102.
117. Claudia Ovando, *Diego Rivera: El agua, origen de la vida*, 18.
118. Quoted in Daniel Vargas Parra, "Apuntes para la iconología de un mural," 74.
119. Herrera, "Una vida dedicada a crear la vida," 7.
120. Herrera, "Una vida dedicada," 7.
121. Cleaves et al. "Introduction," 4.

CHAPTER 3

1. Marilyn Grace Miller, *Rise and Fall of the Cosmic Race: The Cult of Mestizaje in Latin America*, 44.
2. José Gaos, *Obras completas: Tomo VII*, 538; Samuel Ramos, *Historia de la filosofía en México*, 144–45.
3. Guillermo Hurtado, *La revolución creadora*, xxii, see also footnote 16 in the introduction; Roberto Fernández Retamar, *Pensamiento de nuestra América: Autorreflexiones y propuestas*, 49–51.
4. For example, Raúl Forner Betancourt barely deemed Vasconcelos's project of developing a botanical ethics an "odd suggestion" (curiosa sugestión) or an "extremely personal Vasconcelian intention" ("personalísimo intento vasconceliano") which ends up being "absurd" ("absurdo"). See Forner Betancourt, "El pensamiento filosófico de José Vasconcelos," 168. More recently, Guillermo

Hurtado dismissed in similar terms Vasconcelos's engagement with nonhuman ethics, as I will show later in this chapter.
5. Brais Outes-León, "Energía, termodinámica, y el imaginario tecnológico en 'La raza cósmica' de José Vasconcelos," 270; Antebi, *Embodied Archive*, 96–97.
6. Abelardo Villegas, *El pensamiento mexicano en el siglo XX*, 46.
7. Hurtado, *La revolución creadora*, 74–79, 156.
8. José Vasconcelos, "Bergson en México," 137–40. "Bergson en México" is a lecture delivered by Vasconcelos at a conference devoted to Bergson held in Mexico City in 1941. In *La revolución creadora: Antonio Caso y José Vasconcelos en la Revolución Mexicana*, Guillermo Hurtado traces the fundamental influence of Bergson's philosophy in Caso and Vasconcelos. See also Ignacio Sánchez Prado, "El mestizaje en el corazón de la utopía: La raza cósmica entre Aztlán y América Latina." Sánchez Prado suggests that the idea of "spirit" in Vasconcelos is the "conceptual product" of two sources: the neoplatonic notion of the One (Plotinus) and Bergson's "vital impulse."
9. Vasconcelos, "Bergson en México," 140.
10. Villegas, *El pensamiento mexicano en el siglo XX*, 40–41.
11. Vasconcelos, *La revulsión de la energía*, 12.
12. Vasconcelos, *La revulsión de la energía*, 21.
13. Hurtado, *El pensamiento del segundo Vasconcelos*, 46.
14. Vasconcelos, *La revulsión de la energía*, 3.
15. Vasconcelos, *La revulsión de la energía*, 4.
16. Laura Torres-Rodríguez, *Orientaciones transpacíficas: La modernidad mexicana y el espectro de Asia*, 119–29. Torres-Rodríguez traces Vasconcelos's engagement with teosophical ideas.
17. Vasconcelos, *La revulsión de la energía*, 20.
18. Vasconcelos, *Tratado de metafísica*, 221.
19. Vasconcelos, *La revulsión de la energía*, 5.
20. Vasconcelos, *La revulsión de la energía*, 9.
21. Vasconcelos, *Tratado de metafísica*, 62.
22. Vasconcelos, *La revulsión de la energía*, 11.
23. Vasconcelos, *La revulsión de la energía*, 11.
24. Vasconcelos, *Tratado de metafísica*, 205.
25. Vasconcelos, *Tratado de metafísica*, 205–6.
26. Hurtado, *El pensamiento del segundo Vasconcelos*, 43–44.
27. Hurtado, *El pensamiento del segundo Vasconcelos*, 51–52.
28. One rare exception is Margarita Vera Cuspinera, *El pensamiento filosófico de Vasconcelos*, 133–37.
29. Dalton, *Mestizo Modernity*, 31–58.
30. Antebi, *Embodied Archive*, 99.
31. Vasconcelos, *Ética*, 355, 382.
32. Vasconcelos, *Ética*, 382–83.

33. Vasconcelos, *Ética*, 383.
34. Vasconcelos, *Ética*, 116.
35. Vera Cuspinera, *El pensamiento filosófico de Vasconcelos*, 135–36.
36. Vasconcelos, *Ética*, 369.
37. Vasconcelos, *Ética*, 366.
38. Vasconcelos, *Ética*, 367.
39. Vasconcelos, *Ética*, 368; Vasconcelos, "Caballos: Velocidad," 178.
40. Vasconcelos, "Caballos: Velocidad," 171.
41. Vasconcelos, "Caballos: Velocidad," 177.
42. Vasconcelos, "Caballos: Velocidad," 178.
43. Vasconcelos, "Caballos: Velocidad," 171.
44. Vasconcelos, *Ética*, 369, 372.
45. Vasconcelos, "Caballos: Velocidad," 173.
46. Vasconcelos, "Caballos: Velocidad," 175.
47. See for example Thomas Nagel, "What Is It Like to Be a Bat?" 439.
48. Vera Cuspinera, *El pensamiento filosófico de Vasconcelos*, 135; Vasconcelos, *Ética*, 390–91.
49. Vasconcelos, *Ética*, 270–71.
50. Vasconcelos, *Ética*, 389.
51. Vasconcelos, *Tratado de metafísica*, 270.
52. Vasconcelos, *Ética*, 355. Emphasis mine.
53. I will focus on the work of Jean Henri Fabre and Sir J. Chandra Bose, but Vasconcelos also builds on the research of other scientists such as Jakob von Uexküll, Herbert Spencer Jennings and Edgar Dacque.
54. Luis Garrido, *José Vasconcelos*, 75; Dalton, *Mestizo Modernity*, 35. Vera Cuspinera, *El pensamiento filosófico de Vasconcelos*, 163–65.
55. Dalton, *Mestizo Modernity*, 36.
56. This interpretative framework was advanced by Raquel Tibol's *Frida Kahlo: Una vida abierta* (1983) and Hayden Herrera's *Frida: A Biography of Frida Kahlo* (1983). The commercial success of the latter book contributed immensely both to Kahlo's rise to fame and the establishment of the predominant biographical model of interpretation. The 2002 Hollywood film *Frida*, starring Salma Hayek and directed by Julio Taymor, was adapted from Herrera's biography, contributing even more effectively to the general public's obsession with Kahlo's life. In the last two decades, the commercial use of Kahlo's image has grown exponentially, giving rise to a global phenomenon in pop culture dubbed Fridamania.
57. The bibliography on Kahlo's work is immense, but some good starting points are Sarah Lowe, *Frida Kahlo*; Nancy Deffebach, *María Izquierdo and Frida Kahlo: Challenging Visions in Modern Mexican Art*; and Adriana Zavala, *Becoming Modern, Becoming Tradition: Women, Gender, and Representations in Mexican Art*.

58. For a discussion of Kahlo's garden, see Zavala, "Frida Kahlo: Art, Garden, Life"; for an overview of Kahlo's visit to Burbank's house, see Cecilia Stahr, *Frida in America: The Creative Awakening of a Great Artist*, 138–45.
59. Lowe, *Frida Kahlo*, 80.
60. Nancy Deffebach, "Images of Plants in the Art of María Izquierdo, Frida Kahlo, and Leonora Carrington," 140.
61. Lucretia Hoover Giese, "A Rare Crossing: Frida Kahlo and Luther Burbank," 61–64.
62. Giese, "A Rare Crossing," 58.
63. Giese, "A Rare Crossing," 70.
64. Nancy Deffebach, "Human-Plant Hybrids in the Art of Frida Kahlo and Leonora Carrington: Sources, Context and Issues," 108.
65. Herrera, *Frida: A Biography of Frida Kahlo*, 123.
66. Emanuele Coccia, *The Life of Plants: A Metaphysics of Mixture*, 81.
67. The list includes *El sueño o autorretrato onírico* (1932), *Frida y el aborto* (1932), *Raíces* (1943), and several still-life paintings such as *La flor de la vida* (1944). See Mia D'Avanza and Joanna L. Groarke, "Plates: Artworks in the Exhibition," 58–85. See also Deffebach, "Human-Plant Hybrids in the Art of Frida Kahlo and Leonora Carrington," 101–8.
68. As with most of Kahlo's paintings, *El abrazo de amor* has predominantly received autobiographical readings that highlight how Kahlo is addressing her newfound love for Rivera and her unfulfilled desire of becoming a mother. See, for example, Herrera, *Frida*, 377–78. Similarly, Gannit Ankori suggests that this work "blends biographical references with overt Christian, Hindu, and pre-Columbian symbols." Ankori, *Frida Kahlo*, 144.
69. Ankori, *Frida Kahlo*, 145.
70. Herrera, *Frida*, 328.
71. Ankori, *Frida Kahlo*, 147.
72. Kahlo, *The Diary of Frida Kahlo: An Intimate Portrait*, 248.
73. Kahlo, *The Diary of Frida Kahlo*, 249.
74. Kahlo, *The Diary of Frida Kahlo*, 249–50.
75. Anthony Trewavas, *Plant Behavior and Intelligence*, 12–13.
76. Trewavas, *Plant Behavior*, 16–18; Daniel Chamovitz, *What a Plant Knows: A Field Guide to the Senses of Your Garden and Beyond*, 58–59.
77. Vasconcelos, *Tratado de metafísica*, 221.
78. Vasconcelos, *Tratado de metafísica*, 198.
79. Vasconcelos, *Tratado de metafísica*, 197–98.
80. Vasconcelos, *Tratado de metafísica*, 199.
81. Vasconcelos, *Tratado de metafísica*, 270.
82. Chamovitz summarizes the results of the latest research regarding plants' consciousness as follows: "Plants are acutely aware of the world around them. They are aware of their visual environment; they differentiate between red, blue,

far-read, and UV lights and respond accordingly. They are aware of aromas surrounding them and respond to minute quantities of volatile compounds wafting in the air. Plants know when they are being touched and can distinguish different touches. They are aware of gravity: they can change their shapes to ensure that shoots grow up and roots grow down. And plants are aware of their past: they remember past infections and the conditions they've weathered and then modify their current physiology based on these movements." Chamovitz, *What a Plant Knows*, 137–38.

83. William Cronon, "The Trouble with Wilderness; or, Getting Back to the Wrong Nature," 72–75.
84. Jason Moore, *Capitalism in the Web of Life: Ecology and the Accumulation of Capital*, 2.
85. Vasconcelos, *Tratado de metafísica*, 274.
86. Vasconcelos, *Ética*, 371.
87. Coccia, *The Life of Plants*, 81.
88. Deleuze, *Difference and Repetition*, 167.
89. Vasconcelos, *Tratado de metafísica*, 39–40.
90. Michael Marder, *Plant-Thinking: A Philosophy of Vegetal Life*, 67.
91. Coccia, *The Life of Plants*, 37.
92. Marder, *Plant-Thinking*, 67–74.
93. Stefano Mancuso, *The Revolutionary Intelligence of Plants: A New Understanding of Plant Intelligence and Behavior*, 177–78.
94. Vasconcelos, *Ética*, 384.
95. Vasconcelos, *Ética*, 384.
96. Vera Cuspinera, *El pensamiento filosófico de Vasconcelos*, 135.
97. Vasconcelos, *Ética*, 389.
98. Vasconcelos, *Ética*, 392.
99. Vasconcelos, *Ética*, 389.
100. Vasconcelos, *Ética*, 393.
101. Vasconcelos, *Ética*, 384.
102. Deleuze and Guattari, *A Thousand Plateaus*, 7–9.
103. Itzhak Bar-Lewaw, "La revista Timón y la colaboración nazi de José Vasconcelos," 152–56.
104. Vasconcelos, *Ética*, 391.
105. Jacques Derrida, *Dissemination*, 52.
106. Sánchez Prado, "El mestizaje en el corazón," 175; Miller, *Rise and Fall of the Cosmic Race*, 30.
107. Sánchez Prado, "El mestizaje en el corazón," 174–76; Palou, *El fracaso del mestizo*, 15; Agustín Palacios, "Multicultural Vasconcelos: The Optimistic, and at Times Willful, Misreading of 'La raza cósmica,'" 418.
108. Silvia Spitta, "Of Brown Buffaloes, Cockroaches, and Others: 'Mestizaje' North and South of the Río Bravo," 336; Susan Antebi, "Prometheus Re-bound: Disability, Contingency and the Aesthetics of Hygiene in Post-Revolutionary

Mexico," 207; Antonio Cornejo Polar, "Mestizaje, Transculturation, Heterogeneity," 116; Joshua Lund, *The Impure Imagination: Toward a Critical Hybridity in Latin American Writing*, 55.

109. Laura Torres-Rodríguez, "Orientalizing Mexico: 'Estudios indostánicos' and the Place of India in José Vasconcelos's 'La raza cósmica,'" 86; Sánchez Prado, "El mestizaje en el corazón," 183.

110. Ana María Alonso, "Conforming Disconformity: 'Mestizaje,' Hybridity, and the Aesthetics of Mexican Nationalism," 464.

111. Nancy Stepan, *The Hour of Eugenics: Race, Gender and Nation in Latin America*, 65.

112. Stepan, *The Hour of Eugenics*, 30–32.

113. Antebi, *Embodied Archive*, 93.

114. Alonso, "Conforming Disconformity," 464.

115. Antebi, *Embodied Archive*, 99.

116. Vasconcelos, *La raza cósmica*, 16.

117. Vasconcelos, *La raza cósmica*, 18, 29.

118. Juan Carlos Grijalva, "Vasconcelos o la búsqueda de la Atlántida: Exotismo, arqueología y utopía del mestizaje en 'La raza cósmica,'" 434; Vasconcelos, *La raza cósmica*, 20.

119. Vasconcelos, *La raza cósmica*, 13.

120. Vasconcelos, *La raza cósmica*, 13.

121. Vera Cuspinera, *El pensamiento filosófico de Vasconcelos*, 135; Vasconcelos, *Ética*, 392.

122. Vasconcelos, *Ética*, 380.

123. Vasconcelos, *Ética*, 392.

124. Torres-Rodríguez, "Orientalizing Mexico," 88–89.

125. Vasconcelos, *Estudios indostánicos*, 191.

126. Vasconcelos, *La raza cósmica*, 25.

127. Vasconcelos, *La raza cósmica*, 20.

128. Williams Flores, *Ecosophies, Technology, and Luddism: An Ecocritical Perspective to the Study of 'The Cosmic Race,' 'Doña Bárbara,' 'One Hundred Years of Solitude,' 'The Storyteller,' and 'Mantra,'* 43.

129. Brais Outes-León, "Energía, termodinámica, y el imaginario tecnológico en 'La raza cósmica' de José Vasconcelos," 62; Jerry Hoeg, *Science, Technology, and Latin American Narrative in the Twentieth Century and Beyond*, 64–65.

130. Vasconcelos, *La raza cósmica*, 224.

131. Moore, *Capitalism in the Web of Life*, 83–85.

132. Coccia, *The Life of Plants*, 81.

133. Vasconcelos, *Tratado de metafísica*, 230.

134. Outes-León, "Energía, termodinámica," 272.

135. Vasconcelos, José. *Tratado de metafísica*, 230.

136. Vasconcelos, *La raza cósmica*, 28.

137. Vasconcelos, *La revulsión de la energía*, 9.

138. Vasconcelos, *La revulsión de la energía*, 17.
139. Vasconcelos, *La revulsión de la energía*, 9.
140. Vasconcelos, *La raza cósmica*, 22.
141. Vasconcelos, *La raza cósmica*, 34.
142. Vasconcelos, *La revulsión de la energía*, 9.
143. Vasconcelos, *Tratado de metafísica*, 270.

CHAPTER 4

1. Dr. Atl was involved in the little-known avant garde group Action d'Art spearheaded by André Colomer and Gérard Lacaze-Duthiers. The latter artist contributed in the 1920s to the journal of L'Association Internationale Biocosmique founded by Herrera and others. It is possible that Lacaze-Duthiers and Dr. Atl may have come in contact with biocosmic ideas since the early days of Action d'Art.
2. For an overview of Dr. Atl's life, see Antonio Luna Arroyo, *El Dr. Atl: Sinopsis de su vida y su pintura*; Olga Sáenz, *El símbolo y la acción: Vida y obra de Gerardo Murillo, Dr. Atl*.
3. For an overview of Naui Olin's life, see Adriana Malvido, *Nahui Olin: La mujer del sol*; and Elena Poniatowska, "Nahui Olin: La que hizo olas."
4. Nahui Olin's collection of books included two novels by Flammarion, *Urania* and *L'Atmosphère*. Furthermore, Dr. Atl's *Un hombre más allá del universo*, analyzed later in this chapter, bears the influence of Flammarion's novels. See Mariano Meza Marroquín, "Nahui Olin y la síntesis de cosmos," in *Nahui Olin: La mirada infinita*, 43–46.
5. Peter Krieger, "Dr. Atl's Geo-graphies: Aesthetic Transformations of Telluric and Atmospheric Energy," 16; Cuauhtémoc Medina, *Olinka: La ciudad ideal del Dr. Atl*, 63.
6. Medina, *Olinka*, 17.
7. Krieger, "Dr. Atl's Geo-graphies," 16.
8. Krieger, "Dr. Atl's Geo-graphies," 15.
9. Krieger, "Dr. Atl's Geo-graphies," 18.
10. Elizabeth Grosz, Kathryn Yusoff, and Nigel Clark. "An Interview with Elizabeth Grosz: Geopower, Inhumanism and the Biopolitical," 132.
11. For an overview of Dr. Atl's mining initiatives, see Antonio Luna Arroyo, *El Dr. Atl*, 81–86.
12. See Dr. Atl, *¡Oro! Más oro: El mundo lo necesita, Méjico puede dárselo* and *Petróleo en el Valle de Méjico: Una golden line en la altiplanicie de Anáhuac*.
13. Dr. Atl, *¡Oro!*, 22.
14. Dr. Atl, *Las sinfonías del Popocatépetl*, 47.
15. See for example Carlos Fonseca, *The Literature of Catastrophe: Nature, Disaster and Revolution in Latin America*, 91–93; and Gerardo Casado Navarro, *Gerardo Murillo: El Dr. Atl*, 84–86.
16. Dr. Atl, *Las sinfonías*, 30.

17. Dr. Atl, *Las sinfonías*, 113–14.
18. Dr. Atl, *Las sinfonías*, 25.
19. Dr. Atl, *Las sinfonías*, 76.
20. Dr. Atl, *Las sinfonías*, 78.
21. Dr. Atl, *Las sinfonías*, 70.
22. Dr. Atl, *Cómo nace y crece un volcán: El Paricutín*, 47.
23. For an evaluation of Dr. Atl's scientific theories, see Eli de Gortari, "La magia de los volcanes: El Dr. Atl y su ciencia."
24. Dr. Atl, *Cómo nace y crece*, 79.
25. Dr. Atl, *Cómo nace y crece*, 133.
26. Dr. Atl, *Cómo nace y crece*, 42.
27. Dr. Atl, *Cómo nace y crece*, 134.
28. Dr. Atl, *Cómo nace y crece*, 134–35.
29. Dr. Atl, *Cómo nace y crece*, 135.
30. In his prologue to Serrano's *Una nueva perspectiva*, Dr. Atl maintains a similar position: "There is no law in nature: all the laws that are intended to govern it have been invented by man. Our laws are simple accidental interpretations of multiform and changing phenomena. The constant evolution of science is proof" ("En la naturaleza no existe ninguna ley: todas las leyes con que se pretende regirla, las ha inventado el hombre. Nuestras leyes son simples interpretaciones accidentales de fenómenos multiformes y cambiantes. La constante evolución de la ciencia es una prueba"). In Luis G. Serrano, *Una nueva perspectiva: La perspectiva curvilínea, Prólogo, aplicaciones y notas del Dr. Atl*, 7.
31. Dr. Atl, *Cómo nace y crece*, 123.
32. Fonseca, *The Literature of Catastrophe*, 94; Dr. Atl, *Cómo nace y crece*, 78.
33. Dr. Atl, *Las sinfonías*, 59.
34. Dr. Atl, *Las sinfonías*, 104.
35. Dr. Atl, *Las sinfonías*, 47.
36. Dr. Atl, *Las sinfonías*, 113.
37. Renato González Mello, "El Dr. Atl, los arco iris y los fabricantes de células," in *Vanguardia en México: 1915–1940*, 93.
38. See, for example, Sáenz, *El símbolo y la acción*, 379–80; and Amaya Larucea Garritz, *País y paisaje: Dos invenciones del siglo XIX mexicano*, 181–82.
39. Carlos Ashida, "La mirada divina," in *Dr. Atl: Rotación cósmica a cincuenta años de su muerte*, 101.
40. Fausto Ramírez, "Velasco y el Valle de México (1873–1908): Momento narrativo y retórica visual," in *Estética del paisaje en las Américas*, 25–32.
41. Olga Sáenz, *El símbolo y la acción: Vida y obra de Gerardo Murillo*, 488.
42. Serrano, *Una nueva perspectiva*, 11.
43. Serrano, *Una nueva perspectiva*, 10.
44. By the end of his life, once he had one leg amputated and couldn't hike or climb anymore, Dr. Atl also experimented with other kinds of perspectives such as the so-called *aeropaisaje* (aerolandscape), which consisted of creating works

from an aerial perspective provided by an airplane. The aerolandscape, like the curvilinear perspective, constituted a monumental and "universal" approach to landscape painting.
45. Krieger, "Dr. Atl's Geo-graphies," 21.
46. Dr. Atl, *El paisaje*, 12.
47. Rebeca Barquera, *La energía cósmica de la materia: La tecnología del atlcolor en La sombra del Popocatépetl del Dr. Atl*, 11–15.
48. Dr. Atl, *El paisaje*, 13.
49. Jussi Parikka, *A Geology of Media*, 4.
50. Krieger, "Dr. Atl's Geo-graphies," 26.
51. Dr. Atl, *Gentes profanas en el convento*, 71.
52. Barquera, *La energía cósmica*, 11–15. Barquera talks about the "stratigraphy of Atl colors."
53. Poniatowska, "Nahui Olin," 55–56.
54. See for instance Elissa Rashkin and Viviane Mahieux, "La voluntad de escribir: Mujeres en el campo de las letras (1910–1940)," as well as Carolina Narváez, "Nahui Olin: El cuerpo en el verso. La escritura libre de Carmen Mondragón Valseca."
55. Rashkin and Mahieux. "La voluntad de escribir," 418.
56. Meza Marroquín, "Nahui Olin," 43.
57. Patricia López Lopátegui has compiled Nahui Olin's entire literary production, as well as useful bibliographic and hemerographic documentation, in *Nahui Olin: Sin principio ni fin. vida, obra y varia invención*.
58. Rashkin and Mahieux. "La voluntad de escribir," 418–19.
59. Rebeca Julieta Barquera Guzmán and Mariana Rubio de los Santos, "Los colores intangibles de la atmósfera: La plástica de Nahui Olin," in *Nahui Olin: La mirada infinita*, 90.
60. Tomás Zurián, "Nahui Olin y la 'Energía cósmica,'" in *Nahui Olin: Sin principio ni fin*, 412.
61. Nahui Olin, *Energía cósmica*, in Rosas Lopátegui, ed., *Nahui Olin*, 188.
62. Olin, *Energía cósmica*, 198.
63. Olin, *Energía cósmica*, 197.
64. Olin, *Energía cósmica*, 199.
65. Olin, *Energía cósmica*, 195.
66. Olin, *Energía cósmica*, 199.
67. Olin, *Energía cósmica*, 199–200.
68. Olin, *Energía cósmica*, 203.
69. Olin, *Energía cósmica*, 204.
70. Olin, *Energía cósmica*, 205.
71. Olin, *Energía cósmica*, 204.
72. Barquera Guzmán and Rubio de los Santos, "Los colores intangibles," 90; Olin, *Energía cósmica*, 205.

73. Zurián, "Nahui Olin y la 'Energía cósmica,'" in *Nahui Olin: Sin principio ni fin*, 409; Olin, *Energía cósmica*, 192.
74. Olin, *Óptica cerebral*, in Rosas Lopátegui, ed., *Nahui Olin*, 58.
75. Olin, *Energía cósmica*, 193.
76. Olin, *Energía cósmica*, 193.
77. Olin, *Energía cósmica*, 211.
78. Olin, *Energía cósmica*, 211.
79. Olin, *Óptica cerebral*, 67.
80. Olin, *Energía cósmica*, 192.
81. Zurián, "Nahui Olin: La incontenible pasión por escribir," 27; Olin, *Energía cósmica*, 192. See also Zurián, "Nahui Olin y la 'Energía cósmica,'" 409.
82. Olin, *Energía cósmica*, 193.
83. Olin, *Energía cósmica*, 189.
84. Olin, *Energía cósmica*, 189.
85. Olin, *Energía cósmica*, 189.
86. Helge S. Kragh, *Conceptions of the Cosmos: From Myths to the Accelerating Universe: A History of Cosmology*, 125–63.
87. Nahui Olin, *Energía cósmica*, 201.
88. Nahui Olin, *Energía cósmica*, 201.
89. However, *Un hombre más allá del universo* is a sui generis novel that does not use the literary conventions of the science fiction genre. See Gabriel Trujillo Muñoz, *Biografías del futuro: La ciencia ficción mexicana y sus autores*, 71–72. Cuahutémoc Medina, for his part, considers that Dr. Atl's novel is "more akin to the philosophical dramas of the eighteenth century in Europe than to the modern novel understood as an identitarian story of self-development" ("más afín a los dramas filosóficos del siglo XVIII europeo que a la novela moderna entendida como relato identificatorio del autodesarrollo"). Medina, *Olinka*, 153.
90. Françoise Perus, "La ficción literaria del Dr. Atl," 95.
91. González Mello, "El Dr. Atl," 95–97; Medina, *Olinka*, 158–59.
92. Dr. Atl, *Un hombre más allá del universo*, 13–14.
93. Dr. Atl, *Un hombre*, 22.
94. Dr. Atl, *Un hombre*, 23.
95. Dr. Atl, *Un hombre*, 37.
96. Dr. Atl, *Un hombre*, 33–34.
97. Dr. Atl, *Un hombre*, 31.
98. Dr. Atl, *Un hombre*, 38.
99. Dr. Atl, *Un hombre*, 40.
100. See Ken Croswell, *The Alchemy of the Heavens: Searching for Meaning in the Milky Way*, 24–33; Robert W. Smith, *The Expanding Universe: Astronomy's 'Great Debate' 1900–1931*, 77–90.
101. Dr. Atl, *Un hombre*, 60.
102. Dr. Atl, *Un hombre*, 61.

103. Dr. Atl, *Un hombre*, 100, 102.
104. Zurián, "El Doctor Atl," 60.
105. Robert H. Sanders, *Revealing the Heart of the Galaxy: The Milky Way and its Black Hole*, 67–68.
106. Dr. Atl, *Un hombre*, 64.
107. Dr. Atl, *Un hombre*, 59, 65.
108. Dr. Atl, *2 cartas, una para Joaquín Gallo y otra para Agustín Velázquez Chávez*, 14.
109. Dr. Atl, *Un hombre*, 108.
110. González Mello, "El Dr. Atl," 97.
111. Dr. Atl, *Un hombre*, 78.
112. Dr. Atl, *Un hombre*, 85, 83.
113. Dr. Atl, *Un hombre*, 87.
114. Dr. Atl, *Un hombre*, 91.
115. Medina, *Olinka*, 154.
116. Perus, "La ficción literaria del Dr. Atl," 95.
117. Dr. Atl, *Un hombre*, 91.
118. Dr. Atl, *Un hombre*, 116.
119. Dr. Atl, *Un hombre*, 120–21.
120. Dr. Atl, *Un hombre*, 124.
121. Medina, *Olinka*, 157.
122. Beginning with his 1991 bachelor's degree thesis, Cuauhtémoc Medina has widely studied Dr. Atl's utopian project of Olinka. Recently, Medina published *Olinka: La ciudad ideal del Dr. Atl*, a historical reconstruction and analysis which also compiles Dr. Atl's previously unpublished writings on Olinka.
123. In Medina, *Olinka*, 22.
124. Medina, *Olinka*, 188.
125. Medina, *Olinka*, 37–48.
126. In Medina, *Olinka*, 73–74.
127. Dr. Atl, quoted in Medina, *Olinka*, 188.
128. Dr. Atl, quoted in Medina, *Olinka*, 190.
129. Dr. Atl, quoted in Medina, *Olinka*, 190.
130. Dr. Atl, quoted in Medina, *Olinka*, 189.
131. Dr. Atl, quoted in Medina, *Olinka*, 194.
132. Dr. Atl, quoted in Medina, *Olinka*, 218.
133. Medina, *Olinka*, 158–66.
134. Dr. Atl, quoted in Medina, *Olinka*, 206.
135. Dr. Atl, quoted in Medina, *Olinka*, 199.
136. Dr. Atl, quoted in Medina, *Olinka*, 203. To provide a frame of reference, the Estadio Azteca—built approximately ten years after Dr. Atl's plans and still considered the largest stadium in Mexico—has an official capacity for 87,523 spectators.
137. Dr. Atl, quoted in Medina, *Olinka*, 216.

138. Medina, *Olinka*, 62–63.
139. Dr. Atl, quoted in Medina, *Olinka*, 195.
140. Dr. Atl, quoted in Medina, *Olinka*, 195, 225.

EPILOGUE

1. As Sergio Delgado Moya emphasizes, this mural was not commissioned by the Mexican government and painted in public walls, as were earlier postrevolutionary murals. *Delirious Consumption*, 80–81. Instead, *La marcha de la humanidad* was commissioned by investor Manuel Suárez and painted in the lobby of a tourist hotel in Mexico City, which required spectators to pay a fee to see the mural cycle. According to Delgado Moya, the "tension between an allegiance to revolutionary forces and an overt embedment with the world of commerce" (46) that runs through Siquieros's work seemed to be resolved in favor of commercial culture in his last mural.
2. Folgarait, *So Far from Heaven: David Alfaro Siquieros's* The March of Humanity *and Mexican Revolutionary Politics*, 65.
3. See, for example María de las Mercedes Sierra Kehoe, *David Alfaro Siqueiros. Polyforum Siqueiros: Entre el éxtasis y la reinterpretación*, 183–96.
4. Reproduced in Adrián García Cortés, *Siqueiros, Suárez y el Polyforum: Vidas paralelas sin puralelo*, 190.
5. García Cortés, *Siqueiros, Suárez y el Polyforum*, 186.
6. García Cortés, *Siqueiros, Suárez y el Polyforum*, 189.
7. Jameson, *Archaeologies of the Future*, 389.
8. Jacques Lezra, *On the Nature of Marx's Things: Translation as Necrophilology*, 184.

References

Abensour, Miguel. "Persistent utopia." *Constellations* 15, no. 3 (2008): 406–21.
Agamben, Giorgio. *Homo Sacer: Sovereign Power and Bare Life*. Translated by Daniel Heller-Roazen. Redwood City, CA: Stanford University Press, 1998.
———. *Means without Ends*. Translated by Cesare Casarino y Vincenzo Binetti. Minneapolis: University of Minnesota Press, 2000.
———. *The Open: Man and Animal*. Translated by Kevin Attell. Redwood City, CA: Stanford University Press, 2004.
Aguilar Mora, Jorge. *Una muerte sencilla, justa, eterna: Cultura y guerra durante la Revolución Mexicana*. Mexico City: Ediciones Era, 1990.
Allen, Garland E. "Mechanism, Vitalism and Organicism in Late Nineteenth and Twentieth-Century Biology: The Importance of Historical Context." *Studies in History and Philosophy of Biological and Biomedical Sciences* 36 (2005): 261–83.
Alonso, Ana María. "Conforming Disconformity: 'Mestizaje,' Hybridity, and the Aesthetics of Mexican Nationalism." *Cultural Anthropology* 19, no. 4 (2004): 459–90.
Althusser, Louis. *Philosophy of the Encounter: Later Writings, 1978–87*. Edited by Francois Matheron and Oliver Corpet. Translated by G. M. Goshgarian. New York: Verso, 2006.
Ankori, Gannit. *Frida Kahlo*. London: Reaktion Books, 2013.
Antebi, Susan. *Embodied Archive: Disability in Post-Revolutionary Mexican Cultural Production*. Ann Arbor: University of Michigan Press, 2021.
———. "Prometheus Re-bound: Disability, Contingency and the Aesthetics of Hygiene in Post-Revolutionary Mexico." *Arizona Journal of Hispanic Cultural Studies* 17, (2013), 193–210.
Arteaga Castillo, Belinda. *La escuela racionalista en Yucatán: Una experiencia mexicana de educación anarquista (1915–1923)*. Mexico City: UPN, 2005.

Ashida, Carlos. "La mirada divina." In *Dr. Atl, rotación cósmica a cincuenta años de su muerte*, edited by Carlos Ashida, 101–5. Guadalajara: Instituto Cultural Cabañas, 2016.

Atl, Dr. *Cómo nace y crece un volcán: El Paricutín*. México City: El Colegio Nacional, 2017.

———. *Las sinfonías del Popocatépetl*. México City: Ediciones México Moderno, 1921.

———. *¡Oro! Más oro: El mundo lo necesita, Méjico puede dárselo*. Mexico City: Editorial Botas, 1936.

———. *Petróleo en el Valle de Méjico: Una golden line en la altiplanicie de Anáhuac*. Mexico City: Editorial Polis, 1938.

———. *El paisaje: Un ensayo*. Mexico City: n.p., 1933.

———. *Gentes profanas en el convento*. Mexico City: Editorial Botas, 1950.

———. *Un hombre más allá del universo*. Mexico City: Editorial Botas, 1935.

———. *2 cartas, una para Joaquín Gallo y otra para Agustín Velázquez Chávez*. Mexico: Plaquettes Ars, 1946.

Aullet Bribiesca, Guillermo. "Trascendencia del pensamiento y la obra de Alfonso L. Herrera." *Historia Mexicana* 61, no. 4 (April-June 2012), 1525–81.

Bada, J. L., and Lazcano, A. "Prebiotic Soup—Revisiting the Miller Experiment." *Science* 300, no. 5620 (May 2003), 745–46.

Bantjes, Adrian A. "The Regional Dynamics of Anticlericalism and Defanatization in Revolutionary Mexico." In *Faith and Impiety in Revolutionary Mexico*, edited by Mathew Butler, 111–30. New York: Palgrave Macmillan, 2007.

Barquera, Rebeca. *La energía cósmica de la materia: La tecnología del atlcolor en La sombra del Popocatépetl del Dr. Atl*, Master's thesis, UNAM, 2014.

Barquera Guzmán, Rebeca Julieta, and Mariana Rubio de los Santos. "Los colores intangibles de la atmósfera: La plástica de Nahui Olin." In *Nahui Olin: La mirada infinita*, edited by Claudia A. Herrera Martínez, 73–90. Mexico City: Instituto Nacional de Bellas Artes, Museo Nacional de Arte, 2018.

Bar-Lewaw, Itzhak. "La revista Timón y la colaboración nazi de José Vasconcelos." *Actas del cuarto Congreso Internacional de Hispanistas*, vol. 1, edited by Eugenio de Bustos Tovar, 151–56. Salamanca: Universidad de Salamanca, 1982.

Basave Benítez, Agustín Francisco. *México mestizo: Análisis del nacionalismo mexicano en torno a la mestizofilia de Andrés Molina Enríquez*. Mexico City: Fondo de Cultura Económica, 1992.

Beltrán, Enrique. *Contribución de México a la biología: Pasado, presente y futuro*. Mexico City: Editorial Continental, 1982.

———. *Medio siglo de ciencia mexicana, 1900–1950*. Mexico City: SEP, 1952.

Bennet, Janet. *Vibrant Matter: A Political Ecology of Things*. Durham, NC: Duke University Press, 2020.

Bosteels, Bruno. *La comuna mexicana*. Mexico City: Akal, 2021.

Braidotti, Rosi. *Posthuman Knowledge*. Medford, MA: Polity Press, 2019.

Bürger, Peter. *Theory of the Avant-Aarde*. Translated by Michael Shaw. Minneapolis: University of Minnesota Press, 1984.
Calarco, Matthew. *Thinking through Animals: Identity, Difference, Indistinction*. Redwood City, CA: Stanford University Press, 2015.
Campagna, Claudio, Daniel Guevara, and Bernard Le Boeuf. "De-scenting Extinction: The Promise of De-extinction May Hasten Continuing Extinctions." In "Recreating the Wild: Technology, and the Ethics of Conservation," special issue, *Hastings Center Report* 47, no. 4 (2017): S48–S53.
Casado Navarro, Gerardo. *Gerardo Murillo: El Dr. Atl*. Mexico City: UNAM, 1984.
Caso, Antonio. *La existencia como economía, como desinterés y como caridad*. Mexico City: Ediciones México Moderno, 1919.
Castellanos. Israel. *La plasmogenia*. La Habana: Rambla, Bouza y Cía, 1921.
Chamovitz, Daniel. *What a Plant Knows: A Field Guide to the Senses of Your Garden and Beyond*. New York: Scientific American/Farrar, Straus and Giroux, 2012.
Cleaves, Henderson James, Antonio Lazcano, Ismael Ledesma Mateos, Alicia Negrón-Mendoza, Juli Peretó, and Ervin Silva, eds. *Herrera's 'Plasmogenia' and Other Collected Works: Early Writings on the Experimental Study of the Origin Life*. New York: Springer, 2014.
Coccia, Emanuele. *The Life of Plants: A Metaphysics of Mixture*. Translated by Dylan J. Montanari. Cambridge: Polity, 2019.
Coffey, Mary. *How a Revolutionary Art Became Official Culture: Murals, Museums, and the Mexican State*. Durham, NC: Duke University Press, 2012.
Cohen, Shlomo. "The Ethics of De-Extinction." *Nanoethics* 8, no. 2 (2014): 165–78.
Cohn, Robert T. "Official Nationalism in Mexico: Alfonso Reyes and the Hispanization of High Culture at the Turn of the Century." *Anales de la literatura española contemporánea* 12, no. 1/2 (1998): 99–115.
Comisarenco Mirkin, Dina. "*La creación* by Diego Rivera." *Aurora: Journal of the History of Art* 7 (2006): 35–61.
Cornejo Polar, Antonio. "Mestizaje, Transculturation, Heterogeneity." In *The Latin American Cultural Studies Reader*, edited by Ana del Sarto, Alicia Ríos, and Abril Trigo, 116–19. Durham, NC: Duke University Press, 2004.
Cronon, William. "The Trouble with Wilderness; or, Getting Back to the Wrong Nature." In *Uncommon Ground: Rethinking the Human Place in Nature*, edited by William Cronon, 69–90. New York: W.W. Norton, 1996.
Croswell, Ken. *The Alchemy of the Heavens: Searching for Meaning in the Milky Way*. New York: Anchor Books, 1995.
Cuéllar, Margarito (ed). *El derecho a volar: Alfonso Reyes y la ciencia*. Monterrey: Universidad Autónoma de Nuevo León, 2008.
Dalton, David. *Mestizo Modernity: Race, Technology, and the Body in Postrevolutionary Mexico*. Gainesville: University of Florida Press, 2018.
D'Avanza, Mia, and Joanna L. Groarke. "Plates: Artworks in the Exhibition." In *Frida Kahlo's Garden*, edited by Adriana Zavala, Mia D'Avanza and Joanna L.

Groarke, 58–86. New York: DelMonico Books, New York Botanical Garden, 2015.
Deffebach, Nancy. "Human-Plant Hybrids in the Art of Frida Kahlo and Leonora Carrington: Sources, Context and Issues." *Curare: Espacio crítico para las artes*, no. 11 (Summer 1997): 96–128.
———. "Images of Plants in the Art of María Izquierdo, Frida Kahlo, and Leonora Carrington: Gender, Identity, and Spirituality in the Context of Modern Mexico." PhD dissertation, University of Texas, Austin, 2000.
———. *María Izquierdo and Frida Kahlo: Challenging Visions in Modern Mexican Art*. Austin: University of Texas Press, 2015.
Deleuze, Gilles. *Difference and Repetition*. Translated by Paul Patton. New York: Columbia University Press, 1994.
———. "La inmanencia: una vida . . ." In *Ensayos sobre biopolítica: Excesos de vida*, translated by Fermín Rodríguez, 35–40. Buenos Aires: Paidós, 2007.
———. "Literature and Life." In *Essays Critical and Clinical*, translated by Daniel W. Smith and Michael Greco, 1–6. New York/London: Verso, 1998.
Deleuze, Gilles, and Félix Guattari. *A Thousand Plateaus: Capitalism and Schizophrenia*. Translated by Brian Massumi. Minneapolis: University of Minnesota Press, 1997.
———. *What Is Philosophy?* Translated by Hugh Tomlinson and Graham Burchell. New York: Columbia University Press, 1994.
Delgado Moya, Sergio. *Delirious Consumption: Aesthetics and Consumer Capitalism in Mexico and Brazil*. Austin: University of Texas Press, 2017.
Derrida, Jacques. *Dissemination*. Translated by Barbara Johnson. London: Athlone, 1981.
de Carvalho Ramos, Maurício. *A plasmogenia e a síntese conceitual e artificial do protoplasma*. Sao Paulo: Editora LiberArs, 2016.
Esposito, Roberto. *Bíos: Biopolitics and Philosophy*. Translated by Timothy Campbell. Minneapolis: University of Minnesota Press, 2008.
Etkind, Alexander. "Beyond Eugenics: The Forgotten Scandal of Hybridizing Humans and Apes." *Studies in History and Philosophy of Biological and Biomedical Sciences* 39, no. 2 (2008): 205–10.
Giorgi, Gabriel, and Fermín Rodríguez, eds. *Excesos de vida: Ensayos sobre biopolítica*. Buenos Aires: Paidós, 2007.
Fallaw, Ben. "Varieties of Mexican Revolutionary Anticlericalism: Radicalism, Iconoclasm and Otherwise, 1914–1935." *The Americas* 65, no. 4 (April 2009): 481–509.
Fernández Retamar, Roberto. *Pensamiento de nuestra América: Autorreflexiones y propuestas*. Buenos Aires: CLACSO, 2006.
Fierro, Alfonso. "Modeling the Urban Commune: Collective Housing, Utopian Architecture, and Social Reproduction in the Mexican 1930s." *Mexican Studies/Estudios Mexicanos* 38, no. 2 (2022): 272–99.

Finkelman, Leonard. "De-extinction and the Conception of Species." *Biology and Philosophy* 33, no. 5–6 (2018). https://doi-org.dartmouth.idm.oclc.org/10.1007/s10539-018-9639-x.

Fiori de Lima, Nabylla. "Anarquismo individualista, corpo e natureza: A construção de filosofias da natureza contra-hegemônicas na revista espanhola Estudios (1928–1937)." Anais eletrônicos do 15 Seminário Nacional de História da Ciência e da Tecnologia, Florianópolis, Brazil, November 2016. www.15snhct.sbhc.org.br/resources/anais/12/1471289621_ARQUIVO_Anarquismoindividualista,corpoenaturezaAconstrucaodefilosofiasdanaturezacontra-hegemonicasnarevistaespanholaEstudios (1928–1937).pdf.

Flores, Williams. *Ecosophies, Technology, and Luddism: An Ecocritical Perspective to the Study of 'The Cosmic Race,' 'Doña Bárbara,' 'One Hundred Years of Solitude,' 'The Storyteller,' and 'Mantra.'* PhD diss., University of California-Riverside, 2011. doi://escholarship.org/uc/item/7x06d5zr.

Folgarait, Leonard. *Mural Painting and Social Revolution in Mexico, 1920–1940*. Cambridge: Cambridge University Press, 1998.

———. *So Far From Heaven: David Alfaro Siqueiros's* The March of Humanity *and Mexican Revolutionary Politics*. Cambridge: Cambridge University Press, 1987.

Fonseca, Carlos. *The Literature of Catastrophe: Nature, Disaster and Revolution in Latin America*. New York: Bloomsbury Academic, 2020.

Forner Betancourt, Raúl. "El pensamiento filosófico de José Vasconcelos." *Cuadernos Salmantinos de Filosofía*, no. 9 (1982): 147–77.

Fornoff, Carolyn. "Planetary Poetics of Extinction in Contemporary Mexican Poetry." In *Mexican Literature as World Literature*, edited by Ignacio Sánchez Prado, 231–45. New York: Bloomsbury, 2022.

Foucault, Michel. "Right of Death and Power over Life." In *Biopolitics: A Reader*, edited by Timothy Campbell and Adam Sitze, 41–60. Durham, NC: Duke University Press, 2013.

———. *Society Must Be Defended: Lectures at the College de France, 1975–1976*. Translated by David Macey. New York: Picador, 2003.

Gaitán Rojo, Carmen. "Los apreciables jóvenes del Ateneo de la Juventud." In *El Ateneo de la Juventud y la plástica mexicana*, 19–23. Mexico City: INBA, 2010.

Gallo, Rubén. *Mexican Modernity: The Avant-Garde and the Technological Revolution*. Cambridge, MA: MIT Press, 2005.

Gamble, Christopher N., Joshua S. Hanan, and Thomas Nail. "What Is New Materialism?" *Angelaki: Journal of the Theoretical Humanities* 24, no. 6 (2019): 111–34.

Gamio, Manuel. *Forjando patria (pronacionalismo)*. Mexico City: Porrúa Hermanos, 1916.

Gaos, José. "En torno a la filosofía mexicana." In *Obras completas VIII*. Mexico City: UNAM, 1996.

———. *Obras completes: Tomo VII*. Edited by Fernando Salmerón, UNAM, 1996.

García Cortés, Adrián. *Siqueiros, Suárez y el Polyforum: Vidas paralelas sin paralelo.* México City: Polyforum Siqueiros, 2000.
García Morales, Alfonso. "El Ateneo de México: Crónica e interpretación de un proyecto intelectual." In *La revolución intelectual de la Revolución Mexicana (1900-1940)*, edited by Yanna Hadatty Mora, Norma Lojero Vega and Rafael Mondragón Velázquez, 127-53. Mexico City: UNAM, 2019.
Garrido, Luis. *José Vasconcelos.* Mexico City: UNAM, 1963.
Giese, Lucretia Hoover. "A Rare Crossing: Frida Kahlo and Luther Burbank." *American Art* 5, no. 1 (Spring 2001): 52-73.
Giorgi, Gabriel. *Formas comunes: Animalidad, cultura, biopolítica.* Buenos Aires: Eterna Cadencia, 2014.
González Mello, Renato. "El Dr. Atl, los arco iris y los fabricantes de células." In *Vanguardia en México: 1915-1940*, edited by Renato González Mello and Anthony Stanton, 92-99. Mexico City: MUNAL/INBA, 2013.
———. *La máquina de pintar: Rivera, Orozco y la invención de un lenguaje. Emblemas, trofeos y cadáveres.* Mexico City: UNAM, Instituto de Investigaciones Estéticas, 2008.
González Casanova, Pablo. *Un utopista mexicano.* Mexico City: SEP, 1987.
Gortari, Eli de. *La ciencia en la historia de México.* Mexico City: Grijalbo, 1982.
———. "La magia de los volcanes: El Dr. Atl y su ciencia." In *Dr. Atl: Conciencia y paisaje (1875-1964).* Mexico City: UNAM, INBA, 1985.
Grijalva, Juan Carlos. "Vasconcelos o la búsqueda de la Atlántida: Exotismo, arqueología y utopía del mestizaje en 'La raza cósmica.'" *Revista de Crítica Literaria Latinoamericana* 30, no. 60 (2004): 329-45.
Grosz, Elizabeth. *Chaos, Territory, Art: Deleuze and the Framing of the Earth.* New York: Columbia University Press, 2020.
Grosz, Elizabeth, Kathryn Yusoff, and Nigel Clark. "An Interview with Elizabeth Grosz: Geopower, Inhumanism and the Biopolitical." *Theory, Culture and Society* 34, no. 2-3 (2017): 129-46.
Groys, Boris, ed. *Russian Cosmism.* Cambridge, MA: MIT Press, 2018.
Hadatty Mora, Yanna, Norma Lojero Vega, and Rafael Mondragón Velázquez, eds. *La revolución intelectual de la Revolución mexicana (1900-1940).* Mexico City: UNAM, 2019.
Herrera, Alfonso L. *Biología y plasmogenia.* Mexico City: Herrero Hermanos Sucesores, 1924.
———. "Cómo obtuvo México el modelo del Diplodocus: Fue el antiguo Director de Estudios Biológicos quien lo solicitó." *El Universal*, April 11, 1930.
———. "El origen del pensamiento." *Estudios* 10, no. 106 (1932): 23-27.
———. "Estudios de plasmogenia." *Boletín de la Dirección de Estudios Biológicos* 2, no. 1, (1915): 29-63.
———. *El híbrido del hombre y el mono.* Valencia: Cuadernos de Cultura, 1933.
———. *Guía para visitar el Museo Nacional de Historia Natural.* Mexico City: Talleres Gráficos de la Nación, 1922.

———. "Inauguración de la Dirección de Estudios Biológicos." *Boletín de la Dirección de Estudios Biológicos* 1, no. 1 (1915): 5–14.
———. "La aspiración gigantesca de la plasmogenia." *Nervio* 1, no. 3 (1931): 17–19.
———. "La biología en México durante un siglo." *El Demócrata*, Sept. 27, 1921.
———. *La filosofía etérea*. Buenos Aires: Centro Cultural Partenón Minerva, 1919.
———. "La filosofía etérea." *Alborada* 1, no. 7 (July 1, 1917): 1–2.
———. *La Plasmogenia: Nueva ciencia del origen de la vida*. Valencia: Redacción y administración Luis Morote, 1932.
———. "Los cerebros artificiales." *Crisol: Revista de Crítica* 10, no. 59 (1933): 312–14.
———. "Les musées de l'avenir." In *Memorias de la Sociedad Científica Antonio Alzate*, edited by Rafael Aguilar y Santillán, 221–52. Mexico City: Imprenta del Gobierno Federal en el Exarzobispado, 1895.
———. "Mi labor revolucionaria en la enseñanza." *Crisol: Revista Mensual* 7, no. 13 (January 1935): 55–58.
———. *Murmullos del universo*. Mexico City: Instituto Nacional de Investigaciones sobre Recursos Bióticos, Compañía Editorial Continental, 1982.
———. *Nociones de biología*. Mexico City: Imprenta de la Secretaría de Fomento, 1904.
———. *Una ciencia nueva, la plasmogenia*. Mexico City: Tip. de la Viuda de F. Díaz de León Sucesores, 1911.
———. *Una nueva ciencia: La plasmogenia*. Barcelona: Maucci, 1926.
———. "Una vida dedicada a crear la vida." *Hombre de América* 2, no. 8 (February 1941): 7.
Herrera, Hayden. *Frida: A Biography of Frida Kahlo*. New York: Harper & Row, 1983.
Hoeg, Jerry. *Science, Technology, and Latin American Narrative in the Twentieth Century and Beyond*. Bethlehem, PA: Lehigh University Press, 2000.
Huidobro, Vicente. "La creación pura." In *Las vanguardias latinoamericanas: Textos programáticos y críticos*, edited by Jorge Schwartz, 108–13. Mexico City: Fondo de Cultura Económica, 2002.
Hurtado, Guillermo. *El pensamiento del segundo Vasconcelos*. Mexico City: UNAM, 2020.
———. *La revolución creadora: Antonio Caso y José Vasconcelos en la Revolución Mexicana*. Mexico City: UNAM, 2016.
Jameson, Fredric. *Archaeologies of the Future: The Desire Called Utopia and Other Science Fictions*. New York: Verso, 2005.
Janzen, Janet. *Media, Modernity and Dynamic Plants in Early 20th Century German Culture*. Boston: Brill Rodopi, 2016.
Janzen, Rebecca. *The National Body in Mexican Literature: Collective Challenges to Biopolitical Control*. New York: Palgrave Macmillan, 2015.
Kahlo, Frida. *The Diary of Frida Kahlo: An Intimate Portrait*. New York: Abradale Press, 1995.
Klich, Lynda. "Estridentópolis: Achieving a Post-revolutionary Utopia in Jalapa." *Journal of Decorative and Propaganda Arts*, no. 26 (2010): 102–27.

Knight, Alan. "El utopismo y la Revolución Mexicana." *La revolución cósmica: Utopías, regiones y resultados, México 1910–1940*. Mexico City: Fondo de Cultura Económica, 2015.

Kragh, Helge S. *Conceptions of the Cosmos: From Myths to the Accelerating Universe: A History of Cosmology*. Oxford: Oxford University Press, 2007.

Krauze de Kolteniuc, Rosa. *La filosofía de Antonio Caso*. Mexico City: UNAM, 1985.

Krieger, Peter. "Dr. Atl's Geo-graphies: Aesthetic Transformations of Telluric and Atmospheric Energy." In *Dr. Atl, rotación cósmica a cincuenta años de su muerte*, edited by Carlos Ashida, 12–47. Guadalajara: Instituto Cultural Cabañas, 2016.

Larucea Garritz, Amaya. *País y paisaje: Dos invenciones del siglo XIX mexicano*. Mexico City: UNAM, 2016.

Lazcano, Antonio. "El agua y el origen de la vida." In *El agua, origen de la vida en la Tierra: Diego Rivera y el sistema Lerma*, 80–93. Mexico City: Secretaría del Medio Ambiente, Gobierno del Distrito Federal, Museo de Historia Natural y Cultura ambiental, 2012.

———. "What Is Life? A Brief Historical Overview." *Chemistry and Biodiversity*, no. 5, 2008, 1–15.

Ledesma-Mateos, Ismael. *De Balderas a la Casa del Lago: La institucionalización de la biología en México*. Mexico City: Universidad Autónoma de la Ciudad de México, 2007.

Legrás, Horacio. *Culture and Revolution: Violence, Memory and the Making of Modern Mexico*. Austin: University of Texas Press, 2015.

Levitas, Ruth. *The Concept of Utopia*. Oxford: Peter Lang, 2011.

Lezra, Jacques. *On the Nature of Marx's Things: Translation as Necrophilology*. New York: Fordham University Press, 184.

Lowe, Sarah. *Frida Kahlo*. New York: Universe, 1991.

Luna Arroyo, Antonio. *El Dr. Atl: Sinopsis de su vida y su pintura*. Mexico City: Cvltura, 1952.

Lund, Joshua. *The Impure Imagination: Toward a Critical Hybridity in Latin American Writing*. Minneapolis: University of Minnesota Press, 2006.

Lyman Tower, Sargent. "Three Faces of Utopianism Revisited." *Utopian Studies* 5, no. 1 (1994): 1–37.

Maldonado, Benjamín. *La utopía magonista*. Oaxaca: Colegio de Investigadores en Educación de Oaxaca, 2004.

Malvido, Adriana. *Nahui Olin: La mujer del sol*. Mexico City: Circe, 1993.

Mancuso, Stefano. *The Revolutionary Intelligence of Plants: A New Understanding of Plant Intelligence and Behavior*. New York: Simon & Schuster, 2018.

Marder, Michael. *Plant-Thinking: A Philosophy of Vegetal Life*. New York: Columbia University Press, 2013.

Mary, Albert, and Félix Monier. *Précis de solidarité bio-cosmique*. Conflans-Honorine: Idée libre, 1928.

Medina, Cuauhtémoc. *Olinka: La ciudad ideal del Dr. Atl*. Mexico City: El Colegio Nacional, 2018.
Mesler, Bill, and H. James Cleaves. *A Brief History of Creation: Science and the Search of the Origin of Life*. New York: W.W. Norton, 2015.
Meza Marroquin, Mariano. "Diego Rivera y la experiencia en la URSS." In *Diego Rivera y la experiencia en la URSS*, 17–32. Mexico City: INBA, 2017.
———. "Nahui Olin y la síntesis de cosmos." In *Nahui Olin: La mirada infinita*, edited by Claudia A. Herrera Martínez, 34–51. Mexico City: Instituto Nacional de Bellas Artes, Museo Nacional de Arte, 2018.
Miller, Marilyn Grace. *Rise and Fall of the Cosmic Race: The Cult of Mestizaje in Latin America*. Austin: University of Texas Press, 2004.
Mitchell, Robert. *Bioart and the Vitality of Media*. Seattle: University of Washington Press, 2010.
Moore, Jason. *Capitalism in the Web of Life: Ecology and the Accumulation of Capital*. New York: Verso, 2015.
Moraña, Mabel, and Ignacio Sánchez Prado, eds. *Heridas abiertas: biopolítica y representación en América Latina*. Madrid: Iberoamericana-Vervuert, 2014.
Moreno, Roberto. *La polémica del darwinismo en México*. Mexico City: UNAM, 1984.
Morton, Timothy. *Dark Ecology: For a Logic of Future Coexistence*. New York: Columbia University Press, 2016.
Nagel, Thomas. "What Is It Like to Be a Bat?" *Philosophical Review* 83, no. 4 (1974): 435–50.
Narváez M., Carolina. "Nahui Olin: El cuerpo en el verso. La escritura libre de Carmen Mondragón Valseca." *Debate Feminista*, no. 63 (2022): 77–102.
Nealon, Jeffrey. *Plant Theory: Biopower and Vegetable Life*. Redwood City, CA: Stanford University Press, 2016.
Nemser, Daniel. "Biopolitics in Latin America." In *The Encyclopedia of Postcolonial Studies*, vol. 1, edited by Sangeeta Ray, Henry Schwartz, José Luis Villacañas, Alberto Moreiras, and April Shemak, 178–84. Malden, MA: Wiley-Blackwell, 2016.
Coole, Diana, and Samantha Frost, eds. *New Materialisms: Ontology, Agency, and Politics*. Durham NC: Duke University Press, 2010.
Nieuwland, Ilja. *American Dinosaur Abroad: A Cultural History of Carnegie's Plaster Diplodocus*. Pittsburgh, PA: University of Pittsburgh Press, 2019.
Ocaranza, Fernando. *La tragedia de un rector*. Mexico City: Talleres Linotopográficos Numancia, 1943.
Olin, Nahui. *Nahui Olin: Sin principio ni fin. Vida, obra y varia invención*. Edited by Patricia Rosas Lopátegui. Monterrey: University Autónoma de Nuevo León, 2013.
O'Rourke, Kathryn E. "Gardens and Landscapes of Frida Kahlo's Mexico City." In *Frida Kahlo's Garden*, edited by Adriana Zavala, Mia D'Avanza and Joanna L.

Groarke, 87–104. New York: DelMonico Books, New York Botanical Garden, 2015.

Outes-León, Brais. "Energía, termodinámica, y el imaginario tecnológico en 'La raza cósmica' de José Vasconcelos." *Revista de Estudios Hispánicos* 53, no. 1 (2019): 261–82.

Ovando, Claudia. *Diego Rivera: El agua, origen de la vida*. Mexico City: Consejo Nacional para la Cultura y las Artes, 1999.

Paiva de Andrade, Elaine, Jean Faber, and Luiz Pinguelli Rosa. "A Spontaneous Physics Philosophy on the Concept of Ether throughout the History of Science: Birth, Death and Revival." *Foundations of Science* 18, no. 3 (2013): 559–77.

Palacios, Agustín. "Multicultural Vasconcelos: The Optimistic, and at Times Willful, Misreading of 'La raza cósmica.'" *Latino Studies* 15, no. 4 (2017): 416–38.

Palavicini, Félix F. *¡Castigo! Novela mexicana de 1945*. Mexico City: n.p., 1926.

Palou, Pedro Ángel. *El fracaso del mestizo*. Mexico City: Ariel, 2014.

———. "The Ateneo de la Juventud: The Foundations of Mexican Intellectual Culture." In *A History of Mexican Literature*, edited by Ignacio Sánchez Prado, Anna M. Nogar, José Ramón Ruizánchez, 233–45. New York: Cambridge University Press, 2016.

Paquette, Catha. *At the Crossroads: Diego Rivera and His Patrons at MoMA, Rockefeller Center, and the Palace of the Fine Arts*. Austin: University of Texas Press, 2017.

Parikka, Jussi. *A Geology of Media*. Minneapolis: University of Minnesota Press, 2015.

Penn, Sheldon. "*Visión de Anáhuac* (1519) as Virtual Image: Alfonso Reyes's Bergsonian Aesthetic of Creative Evolution." *Journal of Iberian and Latin American Studies* 21, no. 2 (2015): 127–46.

Pereira, Armando. "D. H. Lawrence: México, la utopía imposible." *Literatura Mexicana* 24, no. 1 (2003): 65–90.

Pérez Tamayo, Ruy. *Historia general de la ciencia en México en el siglo XX*. Mexico City: Fondo de Cultura Económica, 2005.

Perus, Françoise. "La ficción literaria del Dr. Atl." In *Dr. Atl: Conciencia y paisaje (1875–1964)*. Mexico City: UNAM, INBA, 1985.

Pineda Franco, Adela. *The Mexican Revolution on the World Stage: Intellectuals and Film in the Twentieth Century*. Albany, NY: SUNY Press, 2019.

Poniatowska, Elena. "Nahui Olin: La que hizo olas." In *Las siete cabritas*. Mexico City: Era, 2000.

Povinelli, Elizabeth. 2016. *Geontologies: A Requiem to Late Liberalism*. Durham, NC: Duke University Press.

Pratt, Mary Louise. *Planetary Longings*. Durham, NC: Duke University Press, 2022.

Quintana Navarrete, Jorge. "El sueño de Salvador Alvarado: Socialismo utópico y subalternidad en el Yucatán Revolucionario." *Chasqui: Revista de literatura latinoamericana* 48, no. 1 (2019): 166–79.

———. "La utopía a prueba: Formas heterogéneas de vida en 'Eugenia' de Eduardo Urzaiz." *Revista Canadiense de Estudios Hispánicos* 44, no. 2 (2021): 463–84.

———. "Sinofobia y mestizaje eugenésico en la prensa de Sonora." In *Las culturas de la prensa en México, 1880–1930*, edited by Yanna Hadatty Mora and Viviane Mahieux, 463–90. Mexico City: UNAM, Instituto de Investigaciones Filológicas, 2022.

Ramírez, Fausto. "Velasco y el Valle de México (1873–1908): Momento narrativo y retórica visual." In *Estética del paisaje en las Américas*, edited by Louise Noelle and David Wood, 21–50. Mexico City: UNAM/Instituto de Investigaciones Estéticas, 2015.

Ramírez Ulloa, Eliseo. "La simulación en la investigación biológica." In *Obras completas. Tomo III: Ciencia, humanismo y sociedad*, 214–20. Mexico City: El Colegio Nacional, 1988.

Ramos, Samuel. *Historia de la filosofía en México*. Mexico City: UNAM Imprenta Universitaria, 1943.

Rashkin, Elissa, and Viviane Mahieux. "La voluntad de escribir: Mujeres en el campo de las letras (1910–1940)." In *La revolución intelectual de la Revolución Mexicana (1900–1940)*, edited by Yanna Hadatty Mora, Norma Lojero Vega, and Rafael Mondragón Velázquez, 405–29. Mexico City: UNAM, 2019.

Reyes, Alfonso. *Obras completas de Alfonso Reyes, Tomo VIII*. Fondo de Cultura Económica, 1996.

———. *Obras Completas: Vol XI*. Mexico City: Fondo de Cultura Económica, 1997.

Richardson, William. "The Dilemmas of a Communist Artist: Diego Rivera in Moscow, 1927–1928." *Mexican Studies/Estudios Mexicanos* 3, no. 1 (Winter 1987): 49–69.

Rivera, Diego. "El Anfiteatro de la Escuela Nacional Preparatoria." In *Textos de arte reunidos y presentados*, edited by Xavier Moyssen, 82–84. Mexico City: UNAM, 1986.

Rivera Garza, Cristina. *Autobiografía del algodón*. Mexico City: Literatura Random House, 2020.

———. *Escrituras geológicas*. Madrid: Iberoamericana Vervuert, 2022.

Romanell, Patrick. *La formación de la mentalidad mexicana: Panorama actual de la filosofía en México, 1910–1950*. Mexico City: El Colegio de México, 1954.

Roosth, Sophia. *Synthetic: How Life Got Made*. Chicago: University of Chicago Press. 2017.

Rosales, Tomás. *El gobierno de mañana: República social sinárquica*. Mexico City: n.p., 1915.

Rossianov, Kirill. "Beyond Species: Il'ya Ivanov and His Experiments on Cross-Breeding Humans with Anthropoids Apes." *Science in Context* 15, no. 2 (2002): 277–316.

Rouaix, Pastor. "La Dirección de Estudios Biológicos y la obra del Profesor Alfonso L. Herrera." In *Vida y obra de Pastor Rouaix*, edited by Salvador Cruz, 17–58. Mexico City: INAH, 1980.

Rovira Gaspar, Ma. Del Carmen. *Una aproximación a la historia de las ideas filosóficas en México: Siglo XIX y principios del XX. Tomo I*, edited by María del Carmen Rovira Gaspar. Querétaro: Universidad Autónoma de Querétaro, 2010.

———. *Dos utopistas mexicanos del siglo XIX: Francisco Severo Maldonado y Ocampo y Juan Nepomuceno Adorno*. Guanajuato: Universidad de Guanajuato, 2013.

———. "El Ateneo de la Juventud." In *Una aproximación a la historia de las ideas filosóficas en México: Siglo XIX y principios del XX*, edited by Ma. Del Carmen Rovira Gaspar, 879–91. Mexico City: UNAM, 1997.

Ruiz Gutiérrez, Rosaura. *Positivismo y evolución: Introducción del darwinismo en México*. Mexico City: UNAM, 1987.

Sáenz, Olga. *El símbolo y la acción: Vida y obra de Gerardo Murillo, Dr. Atl*. Mexico City: El Colegio Nacional, 2005.

Sánchez, Carlos Alberto, and Robert Eli Sanchez Jr., eds. *Mexican Philosophy in the Twentieth Century*. New York: Oxford University Press, 2017.

Sánchez Prado, Ignacio. "El mestizaje en el corazón de la utopía: La raza cósmica entre Aztlán y América Latina." In *Intermitencias americanistas: Estudios y ensayos escogidos (2004-2010)*, 165–88. Mexico City: UNAM, 2010.

———. "Las reencarnaciones del centauro: *El deslinde* después de los estudios culturales." In *Intermitencias alfonsinas: Estudios y otros textos (2004-2018)*. Monterrey: Universidad Autónoma de Nuevo León, 2019.

———. "The Age of Utopia: Alfonso Reyes, Deep Time and the Critique of Colonial Modernity." *Romance Notes* 53, no. 1 (2013): 93–104.

Sanders, Robert H. *Revealing the Heart of the Galaxy: The Milky Way and its Black Hole*. New York: Cambridge University Press, 2014.

Serrano. Luis G. *Una nueva perspectiva. La perspectiva curvilínea: Prólogo, aplicaciones y notas del Dr. Atl*. México City: Cvltura, 1934.

Sierra Kehoe, María de las Mercedes. *David Alfaro Siqueiros. Polyforum Siqueiros: Entre el éxtasis y la reinterpretación*. Mexico City: UNAM, 2016.

Sloterdijk, Peter. "Reglas para el Parque Humano," July 1999. Translated for *Revista Observaciones Filosóficas*, accessed on Oct. 19, 2023. https://www.observacionesfilosoficas.net/lasreglasparaelparque.html.

Smith, Robert W. *The Expanding Universe: Astronomy's 'Great Debate' 1900-1931*. Cambridge: Cambridge University Press, 1982.

Soto Villlafaña, Adrián. "Diego Rivera: El hombre en el cruce de caminos." In *Muralismo Mexicano 1920-1940: Catálogo razonado II*, edited by Ida Rodríguez Prampolini, 181–84. Mexico City: Fondo de Cultura Económica, 2012.

———. "Diego Rivera: La creación." In *Muralismo Mexicano 1920-1940: Catálogo razonado I*, edited by Ida Rodríguez Prampolini, 18–24. Mexico City: Fondo de Cultura Económica, 2012.

Spitta, Silvia. "Of Brown Buffaloes, Cockroaches, and Others: 'Mestizaje' North and South of the Río Bravo." *Revista de Estudios Hispánicos* 35, no. 2 (2001): 333–46.
Svyatogor, Alexander. "Our Affirmations." In *Russian Cosmism*, edited by Boris Groys, 59–62. Cambridge MA: EFlux, MIT Press, 2018.
Stahr, Cecilia. *Frida in America: The Creative Awakening of a Great Artist*. New York: St. Martin's Press, 2020.
Stepan, Nancy. *The Hour of Eugenics: Race, Gender and Nation in Latin America*. Ithaca, NY: Cornell University Press, 1991.
Suárez y López Guazo, Laura. *Eugenesia y racismo en México*. Mexico City: UNAM, 2005.
Tibol, Raquel. *Frida Kahlo: Una vida abierta*. Mexico City: Editorial Oasis, 1983.
Torres-Rodríguez, Laura. "Orientalizing Mexico: 'Estudios indostánicos' and the Place of India in José Vasconcelos's 'La raza cósmica.'" *Revista Hispánica Moderna* 68, no. 1 (2015): 77–91.
———. *Orientaciones transpacíficas: La modernidad mexicana y el espectro de Asia*. Chapel Hill: University of North Carolina Press, 2019.
Trabulse, Elías. *Historia de la ciencia en México: Estudio y textos. Siglo XIX*. Mexico City: Fondo de Cultura Económica/Conacyt 1992.
Trewavas, Anthony. *Plant Behavior and Intelligence*. Oxford: Oxford University Press, 2014.
Trujillo Muñoz, Gabriel. *Biografías del futuro: La ciencia ficción mexicana y sus autores*. Mexicali: University Autónoma de Baja California, 2000.
Urías Horcasitas, Beatriz. *Historias secretas del racismo en México (1920–1959)*. Mexico City: Tusquets, 2007.
Valles, Patricia. *Del anarquismo a la utopía: La visión revolucionaria de Miguel Mendoza Lopez Schwertfeger*. Guadalajara: University de Guadalajara/Centro Universitario de Ciencias Sociales y Humanidades, 1996.
Valverde Téllez, Emeterio. *Bibliografía filosófica mexicana: Tomo Segundo*. León: Imprenta de Jesús Rodríguez, 1913.
Van Horn Kopka, Karel. "La literatura sonorense en la década de los veinte. Caso concreto: 'Yorem Tamegua.'" *Memoria del XII Simposio de Historia y Antropología*, vol. 1 (1987): 460–72.
Vaughn, Mary Kay, and Stephen E. Lewis, eds. *The Eagle and the Virgin: Nation and Cultural Revolution in Mexico, 1920–1940*. Durham, NC: Duke University Press, 2006.
Vargas Parra, Daniel. "Apuntes para la iconología de un mural." In *El agua, origen de la vida en la Tierra: Diego Rivera y el sistema Lerma*, 60–77. Mexico City: Secretaría del Medio Ambiente, Gobierno del Distrito Federal, Museo de Historia Natural y Cultura ambiental, 2012.
Vasconcelos, José. "Bergson en México." In *Homenaje a Bergson*, 133–55. Mexico City: UNAM, 1941.

———. "Caballos: Velocidad." In *El pesimismo alegre*, 171–79. Mexico City: M. Aguilar, 1931.
———. *El monismo estético*. Mexico City: Cultura, 1918.
———. *Estudios indostánicos*. Madrid: Calleja, 1923.
———. *Ética*. Madrid: M. Aguilar, 1932.
———. *La raza cósmica*. Paris: Agencia mundial de librería, 1925.
———. *La revulsión de la energía: Los ciclos de la fuerza, el cambio y la existencia*. Mexico City: n.p., 1924.
———. "México en 1950." In *¿Qué es la revolución?*, 131–39. Mexico City: Ediciones Botas, 1937.
———. *Tratado de metafísica*. Mexico City: México Joven, 1929.
Vaughan, Mary Kay, and Stephen Lewis. Introduction to *The Eagle and the Virgin: Nation and Cultural Revolution in Mexico, 1920–1940*, edited by Mary Kay Vaughan and Stephen Lewis, 1–20. Durham, NC: Duke University Press, 2006.
Vázquez Enríquez, Emily Celeste. "Planetariedad en 'El mal de la taiga' de Cristina Rivera Garza." *Romance Quarterly* 70, no. 1 (2023): 9–24.
Vázquez Martín, Eduardo. "El viaje del Diplodocus." In *El Chopo año por año, 1975–2010*, edited by Manuel Andrade, 129–31. Mexico City: UNAM, Museo Universitario del Chopo, 2011.
Vera Cuspinera, Margarita. *El pensamiento filosófico de Vasconcelos*. Mexico City: Extemporáneos, 1979.
Vieira, Fátima. "The Concept of Utopia." In *The Cambridge Companion to Utopian Literature*, edited by Gregory Claeys and Royal Holloway, 3–27. Cambridge: Cambridge University Press, 2010.
Villagómez, Adrián. "La biología en el muralismo de Diego Rivera." *Ciencias*, no. 45, (Jan.–March 1997): 26–28.
Villaneda, Alicia. *Justicia y libertad: Juana Belén Gutierrez de Mendoza (1875–1942)*. Mexico City: DEMAC, 1994.
Villegas, Abelardo. *La filosofía de lo mexicano*. Mexico City: Fondo de Cultura Económica, 1960.
———. *El pensamiento mexicano en el siglo XX*. Mexico City: Fondo de Cultura Económica, 1993.
Villoro, Luis. "La cultura mexicana de 1910 a 1960." *Historia Mexicana* 10, no. 2 (1960): 196–219.
Wade, Peter. *Race and Ethnicity in Latin America*. London: Pluto Press, 2010.
Williams, Gareth. *The Mexican Exception: Sovereignty, Police, and Democracy*. New York: Palgrave Macmillan, 2011.
———. *The Other Side of the Popular: Neoliberalism and Subalternity in Latin America*. Durham, NC: Duke University Press, 2002.
Wolfe, Bertrand. *The Fabulous Life of Diego Rivera*. New York: Stein and Day, 1963.
Wolfe, Cary. *Before the Law: Humans and Animals in a Biopolitical Frame*. Chicago: Chicago University Press, 2013.

Worringer, Wilhelm. *Abstraction and Empathy: A Contribution to the Psychology of Style*. Translated by de Michael Bullock. Chicago: Ivan R. Dee, 1997.
Young, George M. *The Russian Cosmists: The Esoteric Futurism of Nikolai Fedorov and His Followers*. New York: Oxford University Press, 2012.
Zavala, Adriana. *Becoming Modern, Becoming Tradition: Women, Gender, and Representation in Mexican Art*. University Park, PA.: Pennsylvania State University Press, 2010.
———. "Frida Kahlo: Art, Garden, Life." In *Frida Kahlo's Garden*, edited by Adriana Zavala, Mia D'Avanza and Joanna L. Groarke, 15–40. New York: DelMonico Books, New York Botanical Garden, 2015.
Zurián, Tomás. "Nahui Olin: La incontenible pasión por escribir." In *Nahui Olin: Sin principio ni fin: Vida, obra y varia invención*, edited by Patricia Rosas Lopátegui, 7–30. Monterrey: University Autónoma de Nuevo León, 2013.
———. "El Doctor Atl, Nahui Olin y el valor de la ciencia." In *Dr. Atl, rotación cósmica a cincuenta años de su muerte*, edited by Carlos Ashida, 59–63. Guadalajara: Instituto Cultural Cabañas, 2016.
———. "Nahui Olin y la 'Energía cósmica'." In *Nahui Olin: Sin principio ni fin: Vida, obra y varia invención*, edited by Patricia Rosas Lopátegui, 405–15. Monterrey: University Autónoma de Nuevo León, 2013.

Index

Note: page numbers in *italic* refer to figures.

Absolute, the, 100, 102, 104, 106, 109–10, 128, 132, 134
Adorno, Juan Nepomuceno, 10, 187n23
aesthetics, 34, 102–3, 133
　avant-garde, 52, 146
Agamben, Giorgio, 14, 34, 188n41
ancient rhetoric, 47–48
animality, 7, 62
animal studies, 68, 98, 107, 109, 194n25
Ankori, Gannit, 116, 201n68
Antebi, Susan, 12, 105, 128, 188n37, 188n39, 189n49
anthropocentrism, 16, 104, 121, 188n41
anticlericalism, 65, 194n15
antiquity, 39, 110
apes, 66
　hybridization of humans and, 7, 61–62, 71, 95
　See also primates
Aristotle, 16, 110
art, 3, 24, 49, 51, 57–58, 80, 87, 155–56, 161
　avant-garde, 50, 55
　in capitalist society, 171
　cosmos and, 10

decorative, 45
history, 91
Mexican, 11
Mexican biocosmists and, 10
nationalist landscape, 9
plasmogeny and, 47–48, 50–51
scientific thought and, 43
visual, 146
　See also bioart; Dr. Atl (Gerardo Murillo); Kahlo, Frida; Mexican muralism; Rivera, Diego; Siqueiros, David Alfaro
artificial insemination, 62–64, 95
Ateneo de la Juventud, 8–9, 43, 46, 78, 186n18
atheism, 63–64, 66
Atl. *See* Dr. Atl (Gerardo Murillo)
avant-garde, 51, 55
　aesthetics, 52, 146
　groups, 91–92, 136, 171
　literary theory, 47
　movements, 47, 58
　muralism as, 180
　strategies, 147
　See also under poetry

227

bare life, 32, 42
Bergson, Henri, 8, 46, 99, 199n8
  vitalism of, 10, 46, 72, 99
Bernard, Claude, 21, 46, 79–80
bioart, 24, 51, 192n75
biocosmic solidarity, 41–42, 116, 191–92n54
biocosmic thought, 4, 6, 10, 14–17, 35, 182
  Adorno and, 187n23
  Kahlo and, 115
  Mexican, 151
  Vasconcelos and, 105
biological control, 14, 68, 70
biological life, 14, 37–39, 61, 76–78, 96
  bare, 32
  ethical behavior of minerals and, 105
  evolution of, 68
  of humans, 15
  national identity and, 5
  overcoming, 103
  selfishness and, 80
  universal life and, 46, 132
  See also zoe
biologism, 78–79
biology, 10, 21–22, 26–27, 38, 65, 86
  Caso and, 78, 80
  Darwinist, 11
  England and, 63
  Herrera's philosophy of, 36
  human, 64, 110
  institutionalization of, 21, 82
  modern, 21, 35
  Reyes and, 43
biopolitical theory, 13, 16
biopolitics, 13–14, 187n34
biopower, 13, 15–17, 184, 188n41
biotechnology, 35, 51, 192n75
Bose, J. Chandra, 75, 117–18, 200n53. *See also* metals
botany, 36, 122, 130
Burbank, Luther, 112–14

capitalism, 89–90, 131, 189n48
  expansion of, 25
  US, 88
capitalist accumulation, 12, 25, 120, 131
Carranza, Venustiano, 21, 25
Caso, Antonio, 3, 8, 83, 85
  Bergson's philosophy and, 199n8
  criticism of Herrera, 7, 78–80, 87, 95
  existencia como economía, como desinterés y como caridad, La, 78–80
  organicism of, 81
  philosophy of, 45, 80, 99
  vitalism of, 72
  vitalist philosophy of, 62, 78
Castellanos, Israel, 22, 34
Catholic Church, 18, 34, 44, 65
cerebrología, 10, 174–75
Chapultepec, 21, 60–61, 64, 93, 193n3
  Cárcamo de Dolores, 93–94
chimpanzees, 64, 66, 68–69
  biological mixing of humans and, 63, 67
Chizhevsky, Alexander, 92, 185n6
clinamen, 77
Coffey, Mary, 12, 89
colpoids, 24, 40–41, 45, 48–49, 75, 96
communism, 89–90
  Soviet, 88
consciousness, 16, 72, 119–20, 135, 144, 157, 183
  of amoebas, 40
  of colpoids, 41
  constructive, 142
  of Earth, 141
  human, 98, 103, 106, 131, 134
  plant, 98, 117–18, 120–21, 123, 134, 201n82
  See also Vasconcelos, José
Contemporáneos (journal), 10, 43, 52
cosmological discourse, 10
  Olin's, 151
cosmology, 9–11, 98, 137, 152, 161–62, 164–67, 177
  scientific, 9, 161, 165

cosmos, the, 3, 6, 8, 10–11, 19, 39, 61, 181, 191
   Atl colors and, 9
   biocosmic solidarity and, 191n54
   creación, La (Rivera) and, 2
   Dr. Atl on, 143, 145–46, 148–49, 151, 162–65, 168–69, 172–73, 175, 177
   Herrera on, 41, 55, 57–59, 71
   in *hombre controlador del universe, El* (Rivera), 90
   humanity and, 10, 24, 92, 137
   in Kahlo's work, 111, 113–14, 116
   material interactions with, 48
   Olin on, 151–62
   Vasconcelos on, 98, 102–3, 132–34
   vitality of, 7, 15, 113, 176, 184
culture, 2, 12, 31–32, 34, 78, 179
   Ateneo de la Juventud and Mexican, 186n18
   commercial, 209n1
   humanist understanding of, 51
   lack of, 82, 85
   Latin American, 186n13
   pop, 200n56
   postrevolutionary, 5–7, 10, 13, 15, 136, 180, 182
   Reyes's understanding of, 47
   Western, 59
Curtis, H. D., 165
curvilinear perspective, 9, 149–50, 206n44

Dalton, David, 12, 105, 197n104
Darwin, Charles, 21, 62, 123
   Origin of Species, 22
   theory of evolution, 36, 61, 64–65
Darwinism, 79, 128–29
   England and, 64, 66
   social, 78, 113
   teleological, 74
de-extinction, 69–70
   Neanderthal, 195n29
Deleuze, Gilles, 42, 49, 76–77, 120, 123, 126

Delfino, Victor, 22, 34
Derrida, Jacques, 123, 127
Dr. Atl (Gerardo Murillo), 3, 5, 9–11, 160, 177, 184
   Action d'Art and, 204n1
   aeropaisaje and, 205–6n44
   Atl colors, 9, 150, 206n52
   Boca de volcán, 149
   cerebrología, 10, 174–75
   Cómo nace y crece un volcán, 9, 137, 142–45
   Cráter y la Vía Láctea, 148, 149
   Crear la fuerza, 171, 173
   futuro de la especie, El, 172, 176
   gold and, 139–40
   hombre más allá del universe, Un, 9, 137, 162–67, 173–75, 204n4, 207n89
   landscape paintings of, 9, 137–38, 146–47, 149–50
   Nazi ideology and, 14, 183
   Olin and, 136–37, 151–52, 156
   Olinka project, 9, 138, 170–71, 173–77, 181, 188n48, 208n122, 208n136
   ¡Oro! Más oro, 139–40
   prologue to *Una nueva perspectiva* (Serrano), 205n30
   on science, 8
   scientific theories of, 205n23
   sinfonías del Popocatépetl, Las, 9, 137, 140–42, 145, 163
   on space exploration, 172–73
   Volcán en la noche estrellada, 147, 148–50
   See also volcanism; volcanoes

education, 12, 21, 26, 28, 36
   defective, 129
   scientific, 65
Einstein, Albert, 137, 161, 164, 195n42
electricity, 94, 152, 168
electromagnetism, 137, 139
England, Howell S., 63–64, 66–67

equilibrium, 152, 159
  chemical, 83
  cosmic, 154
  of universe, 153, 155
Estridentistas, 10, 52, 189n48
ethics, 11, 16, 102, 104, 106, 124, 183
  botanical, 4, 8, 98, 105, 111, 117, 121–23, 126, 130, 133–34, 198n4
  chemical, 7, 59, 62, 73–74, 77–78, 81, 83, 95, 117
  of nature, 72–73, 83
  nonhuman, 98, 104–5, 199n4
  of universal life, 72
  zoological, 110
eugenic rationality, 35, 42
eugenics, 12–13, 34, 128, 188n37
evolution, 22, 38, 41, 64–65, 67–68, 76, 84, 94, 154, 172
  creative, 99
  Darwin's theory of, 36, 61, 64, 68
  humanity's, 170
  of science, 205n30
  teleological, 62, 68, 71, 76, 95
evolutionism, 99, 110

Fabre, Henri, 123–25, 200n53
Fedorov, Nikolai, 91–92, 185n6, 195n33
Flammarion, Camille, 10, 137, 160, 204n4
form of life, 4, 14, 23, 79, 105, 110
  of plants, 119, 123–24
Fornoff, Carolyn, 5–6, 186n13
fossils, 60–61, 193n3
Foucault, Michel, 13–14. *See also* biopolitics

Gallo, Rubén, 10, 58, 167
Gamio, Manuel, 19, 25–26
gender, 5, 12, 111
  dissidence, 151
  identities, 11
geology, 36, 177
geopower, 138, 140
Giese, Lucretia Hoover, 113–14
Goethe, Johann Wolfgang von, 30, 123
gold, 139, 150, 167

González Mello, Renato, 163, 167, 197n103
Grosz, Elizabeth, 49, 138
Guattari, Félix, 42, 76–77, 120, 123, 126

Haeckel, Ernst, 21, 36–37
Herrera, Alfonso L., 3–5, 6–8, 10–11, 21–43, 46, 71–77, 82, 88–89, 98, 100, 141, 186n17, 197n96
  anticlericalism of, 65
  artificial beings of, 44, 48
  Association Internationale Biocosmique and, 204n1
  awareness, 39–41, 119
  Diplodocus fossil and, 60–61
  Dirección de Estudios Biológicos (Directorate of Biological Studies) and, 14, 21–22, 24, 26–29, 35, 64, 82, 87
  ether and, 167, 195n42
  experiments of, 7, 22, 49, 51, 85, 87, 94–95, 184, 192n75
  Filosofía etérea, 22, 62, 191n54
  híbrido del hombre y el mono, El, 62–63, 65–70
  hombre más allá del universe, Un (Dr. Atl) and, 163
  material vitalism and, 81, 95
  "Mi labor revolucionaria en la enseñanza," 63–64
  Murmullos del universo, 7, 22, 24, 51–59
  Nociones de biología, 21–22
  nueva ciencia, Una, 22, 23, 29, 50, 195n33
  Ocaranza on, 83–87
  Oparin and, 198n115
  Ramírez on, 85–87
  supermen, 24, 30, 33, 58, 67
  on universal life, 103–4
  "vida dedicada a crear la vida, Una," 96
  Zoología, 63
  See also biocosmic solidarity; Bose, J. Chandra; Caso, Antonio;

colpoids; ethics: chemical; metals;
Museo Nacional de Historia
Natural; plasmogenic experiments;
plasmogeny; Reyes, Alfonso
heterogeneity, 123, 125, 127, 134
homogenization, 126, 128
Homo sapiens, 65–67, 69, 176
horses
   artificial insemination of, 63
   policemen on, 88
   Vasconcelos on, 107–10, 134
Hugo, Victor, 70
Huidobro, Vicente, 47–49, 58
human, the, 4, 35, 42, 98, 120, 188n41
   literature and, 49
   See also Agamben, Giorgio; *Raza cósmica, La*
human being, 24, 106, 109–10, 120–21, 134, 152
   absolute past of, 67
   consciousness and, 119
   hombre controlador del universe, El (Rivera) and, 89–90
   hybridization of monkey and, 62–64
   national community and, 180
   presumed preeminence of, 68
   zoe and, 34
   See also Agamben, Giorgio
human brain, 54, 157–61, 177
   imitation of, 29–30, 86
   power/potential of, 9, 156, 171, 176
human exceptionalism, 16–17
humanism, 17, 78, 183
human species, 3, 13, 15, 68
   biologically improving, 30
   biopolitical hierarchy within, 69
   future of, 176
   gold and, 139
   racial hierarchy within, 62
   space exploration and, 172–73
Hurtado, Guillermo, 104–5, 199n4, 199n8
hybridization, 7, 61–63, 65–70, 76, 113–14
   cosmic, 117

cultural, 127
plant, 113, 128
hybrids, 62, 66–67
human-plant, 8, 98, 114, 117, 134
   See also Kahlo, Frida

identity, 106, 111, 122, 125–26, 194n25
   between animals and humans, 68, 95
   European, 113
   gender, 11
   mestizo, 8, 12
   Mexican, 8, 147
   national, 4–5, 7, 12, 113, 138
   between plants and humans, 116
indigenismo, 12
   in Rivera's work, 197n104
Indigenous people, 12, 14, 34, 67, 179
industry, 26, 28, 107
inorganic, the, 3, 49, 59, 61, 80, 114, 121
inorganic matter, 3–4, 7–8, 10, 15, 39, 62, 72, 118
   ability to self-organize of, 78
   Bergson's vitalism and, 46
   biological life as clinamen of, 77
   ethical behavior of, 134
   life as evolution of, 84
   as nonlife, 16
   nutrition and, 80
   universal flow and, 104
   universal life and, 24
   vital vibrancy of, 141
International Biocosmic Association (Association Internationale Biocosmique), 3, 41, 185n6
Ivanov, Ilya, 63–64, 66–67, 69–70, 193n7
Iztaccíhuatl, 140, 174

Jameson, Fredric, 18, 182
jungle
   ecosystem, 98
   in *Murmullos del universo* (Herrera), 52
   Vasconcelos on, 124–26, 128–30, 133

Kahlo, Frida, 3, 111–17, 121, 200–201nn56–58
abrazo de amor del Universo, la Tierra (México), yo, Diego y el señor Xólotl, El, 114–17, 201n68
human-plant hybrids in work of, 8, 98, 114, 117, 134
Retrato de Luther Burbank, 111–14
See also plant life
Knight, Alan, 18
Krieger, Peter, 138–39, 150

Landesio, Eugenio, 9, 140
Latin America, 180
biopolitics in, 187n34 (see also mestizaje)
scientific journals in, 22
tropical regions of, 131
Legrás, Horacio, 5, 11, 15, 185n8
Lemaître, Georges, 161, 164
Lenin, Vladimir, 88, 90
lifeforms, 7, 38, 118, 128
literature, 8, 46–47, 49, 51
Mexican, 6, 11
postrevolutionary, 10, 187n20
utopian, 18

Mary, Albert, 3, 22, 191n52
materialism, 10, 81, 163
aleatory, 77, 196n60
new, 16 -17, 76, 183
Maxwell, James Clerk, 137, 140, 195n42
medicine, 12, 26–28, 58, 109
Medina, Cuauhtémoc, 163, 170–71, 174–75, 207n89, 208n122
Mendel, Gregor, 128
mestizaje, 12, 14, 19, 26, 33, 111
as biopolitics, 187n34
construction of, 113
deconstruction of, 9, 133
ideological project of, 127
theories of, 4
metals, 75–77, 117
life of, 74, 76–77, 105
precious, 131

metaphysics, 8, 34, 102
Western, 122, 126
mexicanidad, 113, 138
Mexican muralism, 1–2, 188
Mexican Revolution, 1, 5, 11–12, 18–19, 73
biocosmic thought and, 14, 182
Dr. Atl and, 136
novel of, 10
Meza Marroquín, Mariano, 151, 198n109
minerals, 46, 65, 68, 94, 98, 100, 158
chemical life of, 99, 106
ethical behavior of, 106
ethics of, 105
vascular plants and, 121
modernity, 5, 12, 69
Moore, Jason, 120, 131
morality, 72–73, 110
morphogenesis, 27, 78
Murillo, Gerardo. See Dr. Atl (Gerardo Murillo)
Museo Nacional de Historia Natural (National Museum of Natural History), 21, 24, 26, 36, 60, 193n3
museums, 26, 60–61, 64
biological, 36, 61
of natural history, 36, 61

National Preparatory School (Mexico City), 1, 91
Nazism, 126, 175
nonhuman, the, 42, 120
nonhuman beings, 3–4, 15–16, 105, 182
nonlife, 15–16, 46, 59, 78, 188n41
nutrition, 24, 28, 46, 79–80, 83, 110

Ocaranza, Fernando, 7, 62, 82–87, 95
Olin, Nahui (Carmen Mondragón), 3, 5, 10–11, 136–37, 156–57, 161–62, 170, 173, 176–77, 184, 204nn3–4, 206n57
Energía cósmica, 9, 137, 152–55, 161
materia cerebral, 158, 164
Óptica cerebral, 9, 137, 151–52, 158–60
See also human brain

INDEX   233

Oparin, Alexander, 22, 92–93, 198n113, 198n115
organic, the, 3, 49, 59, 61, 80, 114, 121
origin of life, 88
　Herrera on, 6, 11, 21–22, 58–59, 65, 87, 95
　inorganic, 36, 65
　Oparin-Haldane theory of, 92–93

Palacio de Bellas Artes (Mexico City), 88, 92
Paquette, Catha, 88–89
Perus, Françoise, 162–63
philosophy, 59, 136
　ancient Greek, 161
　Bergson's, 99, 199n8
　of biology, 36
　ethereal, 72, 74, 167
　Federov's, 92
　Herrera on, 27–28, 34, 53
　of history of New World, 98
　Marxist political, 91
　materialist, 58, 72, 87
　Mexican, 8, 11, 105, 186n19, 197n96
　Mexican biocosmists and, 10
　monist, 167
　of nature, 6, 8, 35–36, 87
　positivist, 8
　Vasconcelos on, 102, 111, 122
　vitalist, 62
　Western, 131
　Western political, 126
　yogic, 130
　See also Caso, Antonio; Herrera, Alfonso L.; Vasconcelos, José
plant life, 122
　Bose and, 118
　Kahlo and, 111, 116–17
　Vasconcelos on, 8, 11, 105, 109–10, 117, 131
plants, 4, 16, 109–10, 117–24, 126, 158, 183
　in *El abrazo de amor del Universo, la Tierra (México), yo, Diego y el señor Xólotl* (Kahlo), 114, 116

　behavior of, 122–23, 129, 132, 134
　Bose and, 75
　Burbank and, 112–14
　collectivity of, 98, 117, 134
　consciousness of, 98, 117, 120–21, 123, 134, 201–2n82
　in *creación, La* (Rivera), 2
　ethics of, 98, 104–6 (*see also* ethics: botanical)
　genetically altered, 51
　Herrera on, 62, 65, 68, 71
　in *hombre controlador del universe, El* (Rivera), 89
　human beings and, 8 (*see also* hybrids: human-plant)
　life of, 15, 98–99, 110–11, 121, 123–24
　in *marcha de la humanidad en la tierra y hacia el cosmos, La* (Siqueiros), 179
　Mendel and, 128
　plasmogeny and, 23
　Vasconcelos on, 97–100, 104–6, 109–11, 119, 121, 123, 126–27, 130–35
　See also ethics: botanical
plant studies, 16–17, 98
plasmogenesis, 27, 79
plasmogenic experiments, 23–24, 37, 44, 47, 49–51, 71, 79, 83–84, 103
　Herrera's, 59, 87, 192n75
plasmogeny, 6, 10, 21–25, 27–28, 35–36, 41–44, 58, 65, 70, 92–93, 96, 103, 191–92n54
　as artform, 50
　Caso's critique of, 79
　cells created by, 83–84
　eugenics and, 34, 42
　future humanity and, 30–32, 38
　hypothesis of, 76
　Mexican biology and, 82
　poetry and, 48–49, 51
　Ramírez's critique of, 86
　Reyes's critique of, 7, 43, 47–48, 79, 85
　utopianism of, 33
　Valverde Téllez's rejection of, 87

Plato, 30, 72, 99
poetry, 11, 47, 49, 51–55
  avant-garde, 10, 24, 47–49, 166
  humanist conception of, 57
  Mexican contemporary, 5
  modernista, 58
  Olin's, 152
  traditional, 48
  See also Herrera, Alfonso L.:
    Murmullos del universo
politics, 34, 43, 175
Popocatépetl, 140–41, 174
Porfiriato, 19, 26, 53, 78
positivism, 78
  Comte's, 72
  Vasconcelos and, 98
posthumanism, 16–17
postrevolutionary culture, 5, 7, 10, 15, 136, 180
potentiality, 160, 170
  of gold, 139
  of human brain, 156
  of inorganic life, 78
  of nature, 48
Povinelli, Elizabeth, 78, 188n41
primates, 66–67
  proximity between humans and, 62, 69, 106
protobes, 23, 32, 36. *See also* colpoids
protoplasm, 21–23, 27–28, 35, 37–38, 54, 56, 83. *See also* plasmogeny

race, 5, 12, 66, 127–34
  cosmic, 102, 105, 125
  mestizaje and, 9, 111
  See also Raza cósmica, La
racial hierarchy, 12, 62, 66, 127
radioactivity, 152, 156
  brain, 156, 159
Ramírez Ulloa, Eliseo, 82, 85–87
Ramos, Samuel, 97, 186n19
raza cósmica, La (Vasconcelos), 5, 8, 26, 98, 127–34, 189n50
  eugenics and, 34

utopian program of, 117
religion, 34, 44, 64–65, 67, 74, 176
Reyes, Alfonso, 3, 8, 46, 78
"Amiba artificial," 7, 24, 43–47, 58
Rivera, Diego, 2–3, 62, 91–92, 94–95, 198nn108–9
  abrazo de amor del Universo, la Tierra (México), yo, Diego y el señor Xólotl, El (Kahlo) and, 115, 201n68
  agua, origen de la vida en la Tierra, El, 93–94
  Alegoría de California, 112
  creación, La, 1–3, 5, 88, 90–92, 94
  death of, 180
  hombre controlador del universe, El, 7, 88–91, 94, 114
  indigenismo in work of, 197n104
  marriage to Kahlo, 111
  Olin and, 136
  Oparin and, 198n113
Rivera Garza, Cristina, 6, 186n13
Russian cosmism, 91–92, 185n6
Russian Revolution, 63, 91–92. *See also* Soviet Union

science, 3, 8, 59, 65, 103, 161, 171, 177, 179
  of astronautics, 174
  biology as modern, 38
  evolution of, 205n30
  experimental, 30
  Herrera on, 27, 53–54, 70, 73, 95
  in *hombre más allá del universe, Un* (Dr. Atl), 162–63
  marcha de la humanidad en la tierra y hacia el cosmos, La (Siqueiros) and, 181–82
  Mexican, 6, 11, 85, 87
  Mexican biocosmists and, 10
  modern, 89, 102
  Olin on, 154–56, 159
  Reyes on, 43, 49
  of space exploration, 160, 173

Vasconcelos on, 111
  See also biology; plasmogeny
science fiction, 160, 162, 207n89. *See
  also* Dr. Atl (Gerardo Murillo):
  hombre más allá del universe, Un
Serrano, Luis G., 149–50
  nueva perspectiva, Una, 205n30
Shapley, Harlow, 137, 164–65, 167
Siqueiros, David Alfaro, 3, 180–81
  *marcha de la humanidad en la tierra
  y hacia el cosmos, La*, 179–82
Sonora, 26, 187n35, 189n49
sovereignty, 13
  of Catholic Church, 44
  state, 12, 180
Soviet Union, 63, 65, 173
  Bolshevik authorities, 63–64
  Rivera in, 91–92, 198n109
space exploration, 3, 38, 132, 137, 160,
  162–63, 175–75, 180–81. *See also*
  Dr. Atl (Gerardo Murillo); Olin,
  Nahui (Carmen Mondragón)
spiritualism, 46, 78
Sviatogor, Aleksandr, 92

textuality, 11, 15
Tsiolkovsky, Konstantin, 92, 160,
  185n6

universal life, 71, 74, 81, 101, 167–68
  biocosmic thought and, 187n23
  definition of, 84–85
  ethics of, 72
  Herrera on, 104, 116
  Herrera's poetry and, 89
  in Kahlo's work, 113–14, 116–17
  plasmogeny and, 191n54
  in Rivera's work, 88, 90–91, 95
  space exploration and, 181
  Vasconcelos on, 98, 101, 104, 132–33
  vegetal life as, 123
  See also Dr. Atl (Gerardo Murillo)
Urzaiz, Eduardo, 19, 189n49
utopia, 183

Homunculus theory of, 30
  of *Mi sueño* (Alvarado), 189n49
  persistence of, 19
  of *Raza cósmica, La* (Vasconcelos), 9,
    105, 117, 128, 131, 133–34
  technological, 31
utopianism, 18
  Latin American, 190n50
  of plasmogeny, 32
  technological, 7, 10, 90

Valley of Mexico, 9, 140, 147, 174
Valverde Téllez, Emeterio, 87, 197n96
Van Drunen, James, 74, 77
Vasconcelos, José, 3, 7–8, 10–11, 14,
  78, 80, 97–98, 111, 134–35, 184,
  200n53
  "Bergson en México," 199n8
  Estética, 98
  Estudios indostánicos, 130
  Ética, 98, 104–7, 124–27, 129
  on hybridization, 113
  "México en 1950," 189n49
  Nazi ideology and, 183
  pesimismo alegre, El, 107–10
  on plant consciousness, 117–23
  political power and, 19
  revulsión de la energía, La, 98–103
  Rivera and, 1, 91
  Timón, 126
  transcendental plan and, 120, 128
  Tratado de metafísica, 98, 103–4
  Ulises criollo, 97
  vegetarianism and, 130
  See also Ateneo de la Juventud;
    ethics; horses; plant life;
    positivism; *Raza cósmica, La*;
    universal life
vegetal life, 109–11, 113, 118–19, 121–25,
  131, 133–34
Velasco, José María, 9, 138, 140,
  147–48
Villoro, Luis, 8, 18n19
vital function, 22–23, 79–81, 83, 85

vitalism, 10, 35, 43, 45, 62, 78
  Bergson's, 10, 46, 72, 99
  material, 76, 81, 87, 95–96
vitality, 3–4, 7, 23, 45, 90, 115
  Caso's organicism and, 81
  cosmic, 1, 3, 9, 90–91, 94, 152–53, 156–57, 162, 181–82 (*see also* Olin, Nahui)
  of cosmos, 113, 152–55, 184
  of Earth, 137, 140–41, 144, 150–51, 176
  the inorganic and, 77
  of matter, 83
  of metals, 76
  of nonhuman beings, 15, 182

volcanism, 4, 9, 140, 151, 163, 175
volcanoes, 52, 137–42, 144–45, 147, 149, 151, 174–75, 179. *See also* Dr. Atl (Gerardo Murillo); Iztaccíhuatl; Popocatépetl

Wells, H. G., 159–60
Western intellectual tradition, 4, 16, 45
Western philosophical tradition, 72, 77, 104
Western thought, 32, 34–35, 42, 49

Zavala, Adriana, 12, 200–201nn57–58
zoe, 14, 34
zoology, 36, 109–10, 113

www.ingramcontent.com/pod-product-compliance
Lightning Source LLC
Chambersburg PA
CBHW030539230426
43665CB00010B/958